Tribology for Industry 4.0

Tribology for Industry 4.0: Solutions for Surface Engineering explores how data analytics, AI, IoT, and smart sensors can be combined to maximize tribological performance in modern industrial systems. It further examines the use of ML and big data analytics to enhance wear management, friction control, and surface engineering efficiency.

Covering adaptive coatings, smart lubrication systems, and predictive maintenance techniques, the book explains how they dynamically respond to operating conditions, minimizing downtime and maximizing efficiency. It emphasizes eco-friendly tribology techniques to help industries comply with environmental regulations and lessen their carbon footprint, such as cutting energy use, improving lubrication, and implementing sustainable materials. Real-world examples and case studies are included to demonstrate successful implementation in manufacturing, automotive, aerospace, and energy sectors.

Researchers and graduate students studying surface engineering, lubrication, and wear management will benefit from this book.

Digvijay G. Bhosale is an associate professor at the Dr. D. Y. Patil Institute of Technology, India. He is an accomplished academic and chartered professional with over 15 years of experience in research, industry, and teaching. He holds a PhD in mechanical engineering from Veermata Jijabai Technological Institute (VJTI), Mumbai, India, and a master of technology (MTech) from Veermata Jijabai Technological Institute (VJTI), Mumbai, India. His research interests span surface engineering, machine learning, tribology, and material processing. Dr. Digvijay has secured funding from multiple organizations to support his research initiatives. He regularly reviews research papers for surface engineering, tribology and manufacturing journals and has published more than 25 papers in reputed journals.

Hitesh Vasudev is a professor in the School of Mechanical Engineering and head of the Department of Research Impact and Outcome, Research and Development Cell at Lovely Professional University, India. He obtained his PhD from Guru Nanak Dev Engineering College, India. His primary research interests are surface engineering, with particular emphasis on thermal spray technologies; development of nanostructured, multimodal, and high-entropy alloy coatings for high-temperature oxidation and corrosion resistance; thermal barrier coatings (TBCs); and microwave processing of materials. Dr. Vasudev has guided numerous undergraduate and postgraduate students, with 13 PhD scholars successfully awarded degrees under his supervision. He has authored over 250 publications across journals, conferences, and book chapters; co-edited more than 10 books; and holds one granted patent. Dr. Vasudev is a member of the American Society of Mechanical Engineers (ASME).

Advances in Surface Engineering and Tribology

Series Editors: Hitesh Vasudev, Jashanpreet Singh, and Jeyaprakash Natarajan

Advances in Surface Engineering and Tribology explores the interdisciplinary aspects of surface science and engineering, as well as tribology discussions pertaining to the examination of friction, wear, and lubrication. It covers topics ranging from surface modification techniques, nanomaterials and their applications in nanotechnology, catalytic and photocatalytic processes, synthesis of materials, colloids and surfaces, thin film deposition processes, methods for characterizing surfaces, and surface engineering for diverse industrial purposes. The volumes comprehensively explore the complex mechanisms that regulate surface phenomena, presenting both established principles and emerging developments in the field of surface science and engineering. The series offers a forum for professionals, engineers, and researchers in the energy, automotive, aerospace, and biomedical sectors to delve into innovative developments in these interrelated disciplines.

Should you have an idea for this series, please get in contact with Kyra Lindholm (kyra.lindholm@taylorandfrancis.com).

Thermal Spray Coatings for High-Temperature Conditions
Edited by Gaurav Prashar, Hitesh Vasudev, and Lalit Thakur

Microwave Claddings
Engineering Applications

Edited by Sarbjeet Kaushal

Fabrication of Wear and Corrosion Resistant Materials
Surface Processing Techniques

Edited by Gurbhej Singh and Hitesh Vasudev

Metal Oxides
Advances in Sustainability and Nanotechnology

Edited by Virat Khanna, Rohit Jasrotia, Tejeet Singh, and Gulab Singh

Multifunctional and Smart Materials
Advances in Sustainable Engineering Applications

Edited by Hitesh Vasudev and Gurbhej Singh

Microwave Processing of Metallic Materials
Revolutionizing Surface Engineering for Green Manufacturing

Edited by Hitesh Vasudev, Gurbhej Singh, and Amit Bansal

Slurry Erosion
Flow Phenomenon, Complexities, and Protection Methods

Edited by Jashanpreet Singh and Satish Kumar

Artificial Intelligence in Tribology
Edited by Jashanpreet Singh and Hitesh Vasudev

Thermal Spray Technology for Advanced Composite Materials
Edited by Hitesh Vasudev, Gaurav Prashar, and Lalit Thakur

For more information about this series, please visit: www.routledge.com/Adanced-Materials-Processing-and-Manufacturing/book-series/CRCAMPM

Tribology for Industry 4.0

Solutions for Surface Engineering

Edited by
Digvijay G. Bhosale
and Hitesh Vasudev

CRC Press
Taylor & Francis Group
Boca Raton London New York

CRC Press is an imprint of the
Taylor & Francis Group, an **informa** business

Designed cover image: Shutterstock

First edition published 2027
by CRC Press
2385 NW Executive Center Drive, Suite 320, Boca Raton FL 33431

and by CRC Press
4 Park Square, Milton Park, Abingdon, Oxon, OX14 4RN

CRC Press is an imprint of Taylor & Francis Group, LLC

ISBN: 978-1-041-24289-5 (hbk)
ISBN: 978-1-041-24292-5 (pbk)
ISBN: 978-1-003-74115-2 (ebk)

DOI: 10.1201/9781003741152

Typeset in Times
by Deanta Global Publishing Services, Chennai, India

Contents

Contributors

Digvijay Bhosale
Mechanical Engineering Department
Dr. D.Y. Patil Institute of Technology, Pune
India

Satyajeet Gaikwad
Citi Bank
United Kingdom

Umesh Ghorpade
Department of Mechanical Engineering
Dnyanshree Institute of Engineering and Technology, Maharashtra
India

Sana Hassan
Independent Researcher
IEEE
United States

Pratik Rajendra Kachare
Department of Mechanical Engineering
Dnyanshree Institute of Engineering and Technology, Maharashtra
India

Gursimran Kaur
Mittal School of Business
Lovely Professional University, Punjab
India

Shivam Kumar
Western Australian School of Mines
Curtin University
Australia

Jürgen Markus Lackner
Institute of Sensors, Photonics and Manufacturing Technologies
Austria

Amrinder Mehta
Research and Development Cell
Lovely Professional University
India

Jeyaprakash Natarajan
Global College of Engineering and Technology
Oman

Pranal Patil
Department of Mechanical Engineering
Dnyanshree Institute of Engineering and Technology
India

Brijesh Prasad
Department of Mechanical Engineering
Graphic Era (Deemed to be University)
India

Gaurav Prashar
University School of Engineering
Rayat Bahra Professional University
Hoshiarpur, Punjab
India

Sanjay W. Rukhande
Department of Mechanical Engineering
Fr. C. Rodrigues Institute of Technology
India

Jashanpreet Singh
Department of Mechanical Engineering
Chandigarh University
India

Shubhangi Suryawanshi
Department of Information Technology
Dr. D.Y. Patil Institute of Technology
India

Hitesh Vasudev
Research and Development Cell
Lovely Professional University
India

Praveen Kumar Verma
Research and Development Cell
Lovely Professional University
India

1 Introduction to Tribology 4.0 and Surface Engineering in the Smart Manufacturing Era

Digvijay G. Bhosale and Hitesh Vasudev

1.1 INTRODUCTION

Industry 4.0 is changing the paradigm of advanced manufacturing and industrial systems through the integration of cyber-physical systems, the Internet of Things (IoT), advanced analytics, artificial intelligence (AI), and smart sensors. As observed by Ciulli, the Fourth Industrial Revolution is based on the automation of intelligent networks and pervading sensing technologies, and promises a high increase in productivity. For the new operating conditions, the disciplines of tribology (the science of friction, wear, and lubrication) and surface engineering (the designing and modifying of surfaces so that they can withstand degradation) attain a new strategic value. These disciplines used to deal with fixed, static problems in the friction and wear degradation systems, systems of material pairings, and lubricating fluids. These disciplines now must integrate with dynamic intelligent systems that monitor, diagnose, and control friction and wear of machine elements in real time. The economic cost of friction and wear in the world is still considerable. The book *Tribology and Surface Engineering* states that in some countries, wear-related losses alone exceed 6% of the gross domestic product. Moving manufacturing systems to Industry 4.0 architecture will not only reduce this cost but also transform the tribological systems from passive components to smart surfaces and adaptive interfaces in friction and wear systems [1].

The evolution undergone is no longer limited to swap materials or changes in coating thickness; it is the addition of sensors, the application of AI-based predictive systems, and the on-the-fly optimization of surface condition. Surface engineering is at the center of this evolution. The monumental *Surface Engineering: Enhancing Life of Tribological Components* outlines the scope of surface modification technologies, both traditional and modern—such as thermal and laser treatments, ion implantation, and the use of functionally graded materials—aimed at increasing

DOI: 10.1201/9781003741152-1

service life and wear resistance [2]. While these techniques are indispensable, their true potential during Industry 4.0 will be unlocked synergistically. Real-time diagnostics, self-optimizing coatings, and feedback mechanisms that modify surface states and lubrication strategies based on operating conditions will make wear surface engineering "smart." Tribological systems and surface engineering are the focus of unprecedented AI and IoT integration. For instance, Springer's review "Artificial intelligence and edge computing for machine maintenance" discusses how maintenance has shifted from reactive to predictive through edge AI and IoT sensor networks. Integrating sensing and data capture, AI predicts degradation of machine parts and triggers interventions to preempt and avoid machine maintenance. The article encompasses concepts that apply to tribologically active components, including bearings, gears, sliding contacts, and coatings, while also addressing the future of tribology 4.0 [3]. In Figure 1.1, the relationships among friction, wear, lubrication, and surface temperature and their consequences on tool life, material removal rate, and surface integrity during industrial processes are depicted.

The monograph *Coatings Tribology: Properties, Mechanisms, Techniques and Applications in Surface Engineering* continues to be a cornerstone reference, even in the context of recent publications in the area of coatings and surface engineering. It was one of the first to capture a robust methodology for the selection of coatings, covering tribological mechanisms and performance at a detailed industrial level. Although older, its frequent citations are a testimony to the continuous value of meticulously designed surfaces [5]. That said, it is high time for the integration of digital surfaces technology with condition monitoring to be built on this surface technology foundation. This is where sensor-embedded active tribological systems and interfaces come into play. Given the context of Industry 4.0, in combination with other systems, functional surfaces in digital space, with sensing and active responsive systems, can interface as parts of larger functional surface systems.

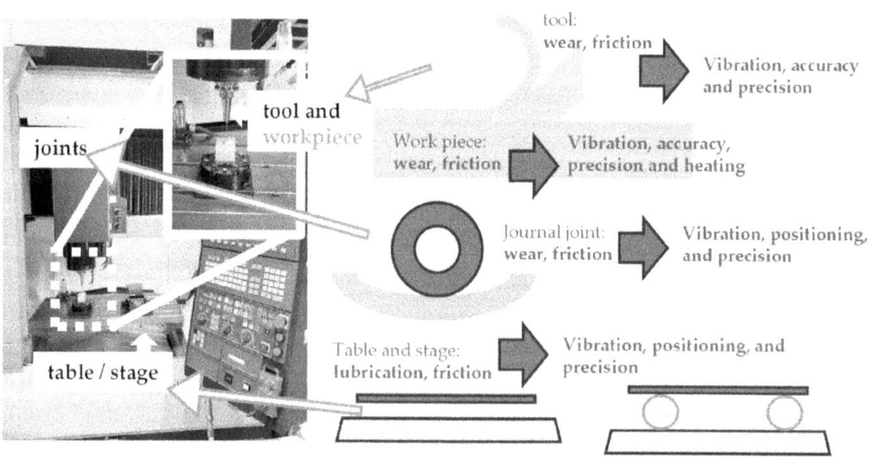

FIGURE 1.1 Tribological factors in machining processes [4]. (CC Attribution 4.0.)

There are opportunities and challenges. From the opportunity side, integrating smart sensing and analytics into tribological systems facilitates sustainable manufacturing by enabling proactive rather than reactive friction and wear management. This means longer-lasting manufacturing components and less lubricants and waste, reduced energy usage, and sustainable manufacturing. The 2019 piece on sustainability, emphasizing biomimetic surface engineering, highlights its importance in heavily utilized, extreme environment sectors like oil and gas. From the challenge side, the integration of surface engineering and tribology with digital systems, AI approaches and IoT frameworks necessitate interdisciplinary teamwork. It also raises the challenges of data standardization and real-time issues of analytics, security, and the reliability of sensors. Furthermore, the surfaces sensing predictive models for tribological failure modes are trained on must remain uncoated, and safe surfaces must remain exposed for sensing [6].

The transition in maintenance strategy—scheduled or preventive maintenance moving to predictive and prescriptive maintenance—brings important considerations for surface engineering and tribology. As outlined in the Springer review on tribology for the years 2020–2021, there is active tribology research in computational modeling, surface diagnostics, and adaptive systems. In the same way, the previous Springer review of tribology discussed the growing research focus in the field on digital and computational technologies. These observations in the literature extend the notion that the integration of tribology, surface engineering, and the technologies of Industry 4.0 is not limited to mere evolution but rather constitutes a shift in paradigm. In this book, we, therefore, characterize Chapter 1 as the foundational gateway that (i) describes the reason why tribology and surface engineering are vital enabling components of the Industry 4.0 ecosystem; (ii) presents the key enabling technologies of IoT, AI, and predictive maintenance, and describes the new ways in which surfaces are engineered, monitored, and maintained; and (iii) outlines the structure of the book to assist the reader in the vision of smart tribological systems and surface engineering for the contemporary digital era. The aim is to assist the reader in transitioning from the more classical concepts of friction, wear, and surface modification to a more advanced, integrated, and intelligent framework in which surfaces are active, sophisticated, and dynamic. This is what the book describes as the "centrality thesis" to understanding the emerging theories of friction, tribology, and the "set" of engineering disciplines associated within the "set" of disciplines networks of "engineering fractals" [7].

1.2 ROLE OF INDUSTRY 4.0 TECHNOLOGIES IN TRIBOLOGY AND SURFACE ENGINEERING

At the core of this shift are three interrelated pillars: IoT, AI and ML (machine learning), and predictive maintenance. Together, they enable a transition from reactive servicing of wear and friction-failure events toward proactive, even prescriptive, management of tribological interfaces and engineered surfaces [8]. Figure 1.2 displays IoT applications integrated with machine learning for real-time monitoring,

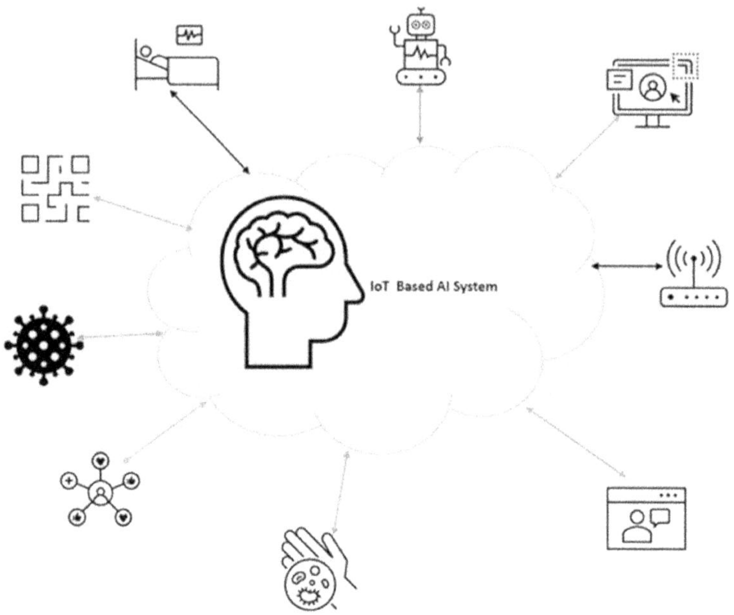

FIGURE 1.2 IoT applications working with ML [9]. (CC Attribution 4.0.)

predictive maintenance, data-driven optimization, and intelligent decision-making across manufacturing, coating processes, and tribological performance enhancement systems.

The first pillar, IoT, equips tribological systems with continuous sensing, connectivity, and real-time data flow. Embedded sensors integrated into machine components such as bearings, gears, or sliding interfaces can capture key tribological parameters—for example, temperature rise, vibration signatures, friction coefficient fluctuations, and wear rate progressions—throughout service life. Smart tribological systems incorporating sensors and actuators can tune themselves online, based on live data of wear, friction, vibration, and temperature. In practical terms, when a bearing's friction coefficient starts to deviate from expected norms, the IoT system can register that change, stream data to a central controller or digital twin, and initiate control responses or flag maintenance requirements. The concept of "internet of surfaces," or "tribotronics," has been introduced in recent literature, signifying how surfaces are being turned into active sensing and feedback elements within an IIoT (industrial IoT) network. In surface engineering, this means coatings or modified surfaces can host sensors or embed functionality that allows direct measurement of tribological state, enabling surface health monitoring rather than simply post-failure inspection [10].

Tribology inherently involves multifaceted interactions in contact mechanics, lubrication regimes, temperature/thermodynamics, material microstructure, and chemistry—rendering its datasets complex, multidimensional, and time varying.

As noted in a recent review, AI applications in tribology are extensive: the authors define five categories of tribo-system information (input I, system intrinsic S, output O, tribological state Ts, and derived state Ds) and map how regression, classification, clustering, and dimension-reduction methods can solve behavior monitoring, prediction, and optimization tasks. Specific examples are now emerging: for instance, a Springer article demonstrates how ML models (Gaussian process regression, support vector regression) were able to predict both wear rate and coefficient of friction of WC–Co-coated mild steel with extremely high accuracy (R^2 ~0.9999) by correlating sliding distance, load, and material/coating parameters. Another work used ML for online detection of wear and friction behavior of a biomedical-graded stainless steel under lubricated conditions, achieving model accuracies of ~98.9% for high-wear conditions. These outcomes highlight the ability of AI/ML to detect subtle patterns in tribological and surface engineering data—patterns that would be extremely difficult to discern through traditional mechanistic modeling or statistical methods alone [11].

The combination of AI/ML analytics and IoT sensing technologies makes predictive maintenance possible. With traditional paradigms for maintenance, systems stay idle until wear or failure occurs, and scheduled maintenance occurrences depend on arbitrary metrics of time or usage. In Industry 4.0, it becomes possible to predict, using analytics on continuously collected wear sensor data, when systems will approach critical thresholds dominated by wear or failure, and subsequently predictive maintenance can intervene before failure occurs. This results in a maintenance paradigm shift from reactive maintenance to prescriptive maintenance, where scheduled maintenance meets the system in an operational state, serviced surfaces are replaced on a just-in-time basis, and system downtimes are avoided or configured to minimal lengths. The aforementioned online wear-detection ML model is a great example of such a shift, because it allows targeted maintenance on systems to the point of avoiding failure-causing wear in a transition from blanket replacement. This is especially true for components that are expensive to manage in the surface layers, where engineered coatings entail extreme surface treatments, costly machine downtimes, or severe failure consequences [12].

The advent of these technologies signifies a new era in tribology and surface engineering. Rather than viewing friction and wear as losses to be managed ex post, systems can now use embedded sensing, self-diagnosis, and adaptive control in friction surfaces and interfaces. This advances the potential for more sustainable and less costly industrial operations. Integration of IoT sensors, AI-driven analytics, and predictive maintenance will prolong the lives of industrial components, slash lubricant and protective coating expenditures, and reduce energy use and maintenance downtime and costs due to diminished friction losses. Yet, the challenges in deploying these technologies are equally significant. The issues of data standardization, the need for reliable sensors in hostile tribological conditions (high pressure and temperature, aggressive corrosion, and severe wear), the security of IoT networks' data, and the availability of high-quality data for machine learning training are particularly detrimental. As the focus of the current trends review highlights, AI and machine learning can offer great value to tribology, but, unlike in other disciplines, the value remains largely untapped. This stems from the advanced complexity of tribological systems and a relative lack of large, standardized data repositories. The other side of the coin is that embedding smart sensors in surface-engineered coatings or surfaces should not compromise coating performance or integrity, nor should it prompt new failure mechanisms [13].

1.3 EVOLUTION OF TRIBOLOGY IN THE CONTEXT OF INDUSTRY 4.0

Methods in surface engineering, including thermal spraying, laser cladding, ion implantation, and additive coating layers, increased the resistance of systems to wear and corrosion, but systems remained physical and without real-time monitoring or adaptation capabilities. In stark contrast, the Industry 4.0 paradigm shifts to a new vision of what can be termed Tribology 4.0: a digitally integrated and flexible, adaptive, and responsive tribological ecosystem. The integration of cyber-physical systems, embedded command and control systems, real-time monitoring, control of surfaces, lubricants, and sliding interfaces promises unprecedented evolution. The physical contact interface is more than a lubricated geometry and finished surface coating; it is now a connected network of sensing and control commanded to adapt in real time. With the unprecedented evolution of control and monitoring, interfaces of tribological systems can adapt and optimize dynamically. Decision-making can be automated, feedback loops enacted, and surface behavior adapted intelligently [14, 15]. Figure 1.3 is a schematic representation of a mini traction machine employing the spacer layer imaging method (SLIM) to measure lubricant film thickness, friction behavior, and contact conditions under controlled load and speed.

An important enabler for this evolution is the concept of a digital twin—a virtual representation of the tribological system (surfaces, coatings, contact mechanics, loads, and environmental conditions). With the digital twin, engineers are able to simulate the progression of wear, test coating performances under different operating scenarios, and develop surface treatments that are designed and perfected both prior to and during service. By continually updating the virtual model with data from the physical interface (e.g., friction coefficient, temperature, and vibration), the system identifies a slipping condition, predicts a failure, and automates the adjustment of surface treatments, lubrication, and maintenance. The paradigm of "design once and maintenance later" is obsolete, replaced with continuous optimization [17]. The significance from a surface engineering perspective is the digital augmentation of the fundamentals. Methods like thermal spraying, laser cladding, and additive manufacturing of surface layers are no longer isolated fabrication steps but constituent steps in a smart and integrated workflow. For example, a laser-clad surface can be equipped with sensors for monitoring wear, which can be integrated with a predictive algorithm that dictates coating adjustment schedules. Surface manufacturing itself becomes integrated with the smart factory environment of Industry 4.0. Similarly, additive manufacturing of tribological surfaces builds in sensor channels and micro-features, enabling near-real-time control of contact topology, lubrication supply, and

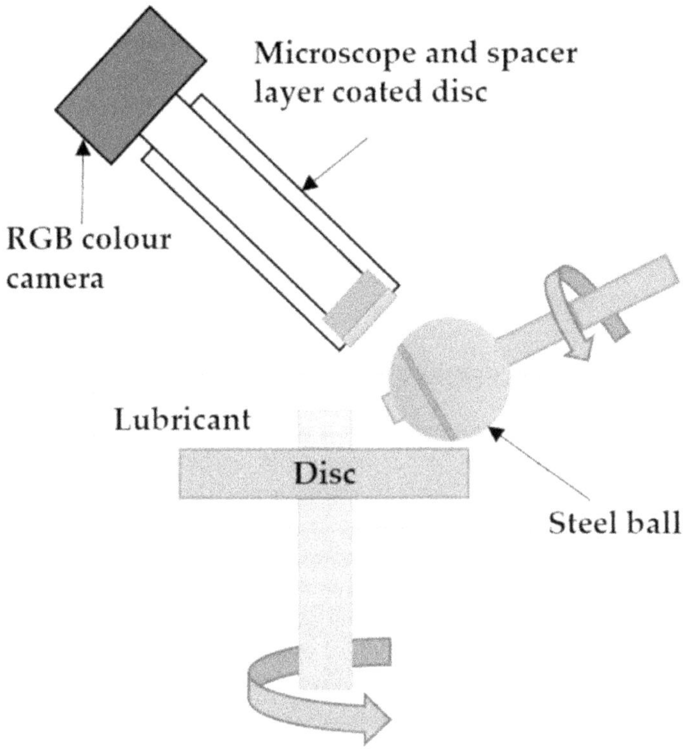

FIGURE 1.3 Schematic representation of mini traction machine, spacer layer imaging (SLIM) test [16].

wear behavior. In effect, the evolution is twofold: first, a shift from purely material/coating centric engineering to system-centric, data-driven tribology; second, a shift from reactive maintenance (replace when worn) to proactive and prescriptive maintenance (predict, adapt, intervene). In smart manufacturing ecosystems, the passage described for wear and friction defines the zone as a monitored metric for optimization rather than a loss. The interfaces being described develop from passive to active components. There are net losses from friction and wear; as opposed to losses, they are viewed as a surface or zone to be monitored and adjusted for optimization. Surface engineering expands from durability to the line: coatings respond to preset conditions, lubrication controls adjust automatically, and surface patterns flex or change in a hybrid/smart manner [18].

Manufacturing is going to be impacted significantly. Systems with digitally überaugmented tribological surfaces provide increased reliability, longer service life, lower maintenance downtimes, and minimized energy losses through optimized friction. They also help achieve the sustainability objectives of Industry 4.0 by reducing lubricants and waste lubricants, failed components, and maintenance footprints. In addition, the new smart tribological systems generate valuable data (sensor streams, wear maps, lifecycle logs) and offer new possibilities for data-driven design of surfaces and coatings. Many elements such as process parameters, layer thicknesses, substrate preparation, microtopography, and even surfaces and coatings design can be optimized for machine learning applications using field data instead of lab data. Other elements should be optimized to improve machine learning applications. Integrating sensors into hostile wear environments without losing surface integrity, contact-mechanical systems data fidelity, complex tribological digital twin construction, surface engineering and data science/AI crossover, and maintenance strategies remodeled around a condition servicing cycle instead of a scheduled servicing cycle are all challenges discovered [19, 20].

1.4 SURFACE ENGINEERING IN SMART MANUFACTURING

As Tribology 4.0 emerges with intelligent materials, adaptive surface technologies, and controlled data processes, it becomes smart surface engineering integrating data-driven processes. Smart surface engineering integrates a wide variety of processes to change the surface qualities of a material, including hardness, friction, corrosion, and wettability to improve and extend a material's performance under service conditions. In smart and adaptive surface engineering, modern surface engineering has evolved a great deal from the passive protection of surfaces. AI, alongside real-time monitoring and additive manufacturing technologies, has made surface engineering pivotal for sustainability in the energy-efficient and intelligent processes of manufacturing anticipated under Industry 4.0.

Historically, static methods such as carburizing, nitriding, electroplating, anodizing, and thermal spraying formed the basis of surface engineering and provided enhanced surface hardness and protection against wear, corrosion, and other surface deterioration. Although the methods provided remarkable advancements, they were deployed with low adaptability, self-regulating systems, or automata. This development has ranged from surfaces being mere protective barriers to fully integrated elements of the mechanical systems as they contribute to the performance and sustainability of the system as a whole [21].

One of the most promising to contribute to this transformation is laser surface texturing (LST). LST employs highly focused laser beams to produce micro- and nanoscale surfaces. The two most relevant differences between LST and any form of laser processing are (i) LST significantly lowers the friction and wear of the surfaces as well as improves the overall durability of the interfaces and (ii) LST produces lubricant wear debris containers. In smart manufacturing, LST is being integrated using AI and machine vision as adaptive surfaces in which the modification is real time. For example, operational data is monitored using laser systems, high-wear areas are diagnosed, and the surface is automatically retextured or repaired during maintenance cycles. This is a major leap in the development of self-adaptive and reconfigurable surface systems where texture geometries and distributions are fixed, evolving with the operating environment to extend service life and optimal performance [22].

Alongside LST, the development of nanocomposite coatings represents another breakthrough in the engineering of smart surfaces. These coatings involve the combination of materials at the nanoscale, which results in exceptional performance attributes. Usually, the hard ceramic matrices of titanium nitride (TiN), chromium nitride (CrN), or diamond-like carbon (DLC) are reinforced with nanoadditives of graphene, carbon nanotubes (CNTs), molybdenum disulfide (MoS_2), or boron nitride (BN). The hard phase with lubricating nanomaterials synergistically provides the coatings with increased toughness, brilliant wear resistance, and ultra-low friction. Furthermore, the modern array of nanocomposite coatings gives the option of incorporating embedded sensing elements monitoring stress, temperature, or wear through piezoelectric or magnetoresistive sensors. These coatings enable mechanical systems to convert modern self-sensing technology, which communicates system status to centralized control, converting mechanical systems into smart systems. The integration of this system with AI, predictive maintenance, and analytics enables proactive maintenance, reducing unplanned downtime and optimizing the performance of the equipment across its lifecycle [23]. Figure 1.4 shows the laser processing setup, which illustrates the beam delivery system, focal optics, and scanning patterns employed for surface texturing, cladding, and microstructural alteration to enhance coating adhesion and wear resistance.

Adaptive surfaces have advanced with the introduction of stimuli-responsive and self-healing coatings. These materials can alter certain surface properties, including the friction coefficient, wettability, and adhesion, when subjected to temperature- and pressure-responsive stimuli as well as pH and electric potentials. Thermoresponsive polymers combined with tribological coatings, for example, can have their viscoelastic properties changed significantly, providing variable

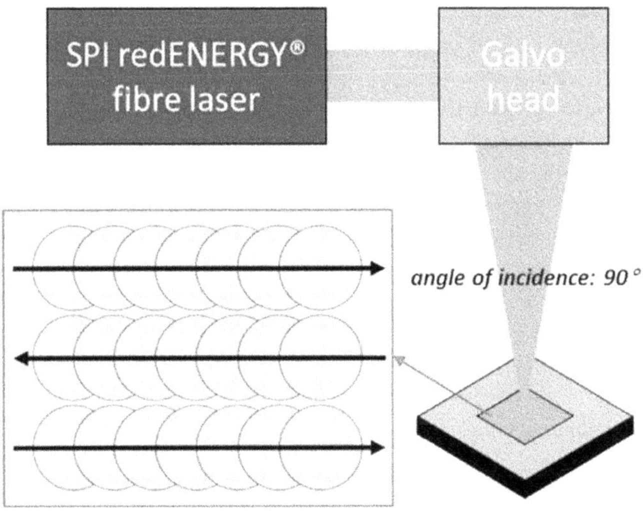

FIGURE 1.4 Laser setup and scanning patterns [23].

adaptive lubrication with the polymer and coating combination to the overlying surface under varying thermal loads. Electroresponsive coatings can similarly provide responsive-coating systems that frictionally interact with surfaces under electric or magnetic fields of varying strength and configuration, permitting control over the surface charge density and hydrophobicity to alter the friction dynamically. In self-healing coatings, microcapsules with healing agents and polymer networks with reversible bonds can autonomously regain microstructural cracks and protective properties that are worn away [24].

The innovation that additive manufacturing (AM) brings to the design and construction of tribological surfaces cannot be overstated. AM can manufacture functionally graded materials (FGMs) and architected surfaces. Such design and construction capabilities cannot be achieved through traditional subtractive methods. AM also offers surfaces that range from base materials through various customizable gradient properties to the surfaces. Engineers using selective laser melting (SLM), directed energy deposition (DED), and binder jetting combined with conventional methods create components with a hardened outer layer that is wear resistant and transitions to a core that is tough and ductile. This configuration reduces the chances of delamination and brittle fracture. AM can also incorporate sensor networks and smart microstructures that monitor and record strain, temperature, and wear. In addition, hyperassisted manufacturing, which combines AM with other surface coating techniques like plasma spraying and laser cladding, is designed to have multiple layers that incorporate systems with controllable porosity, adhesion, determined thermomechanical properties, and porosity. The systems can be integrated with machine learning predicting the interrelationship of the manufacturing process and the resultant performance. This promises a digital craft revolution where AM will become a tool to autonomously balance surface design with performance goals [25, 26].

There have been advances in the field of surface engineering as a result of digital twin technology and AI innovations. These AI capabilities also include surface composition and microstructures as well as process parameters in addition to predicting and analyzing data from experiments, simulations, and in-service sensors. Rapid optimization through experimental limits can be accomplished via machine learning, specifically neural and genetic algorithms, which are increasingly popular for modeling complex relationships, particularly with deposition conditions and tribological performance. Engineers can assess surface wear, degradation, and coating failure performance under theoretical varying operational conditions. Integrated with embedded sensor data, engineers can dynamically assess and control surface processes in real time. Sensors enable closed-loop control, which offers real-time control to surface processes. Predictive modeling offers enhanced assessment of performance throughout operational life. Continuing optimization offers improved surface performance throughout the operational life of the product. Figure 1.5 is a conceptual model of digital twin data flow illustrating integration between physical system, virtual simulation, and real-time sensor feedback for predictive analysis, process optimization, and intelligent coating performance monitoring.

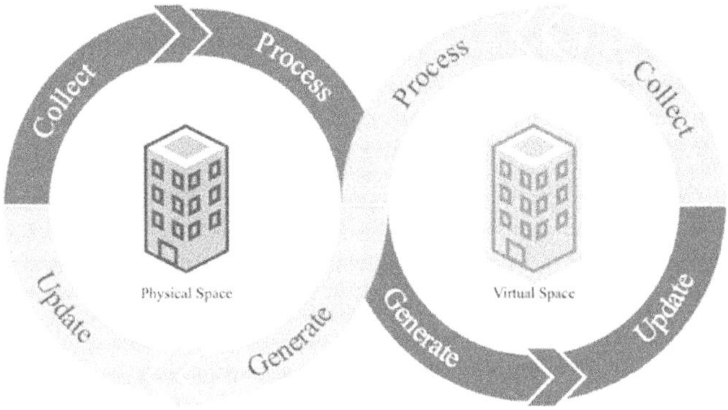

FIGURE 1.5 Digital twin data flow conceptual model [27].

From an ecological standpoint, advanced surface engineering is highly relevant to circular manufacturing. Smart surfaces reduce material waste, extend the life of components, and decrease energy dissipation due to friction. For instance, reconfigurable or reprocessable surfaces can undergo treatments such as local laser remelting and plasma reactivation, as well as microwave-assisted coating rejuvenation, which restoratively refurbish coatings and reallocate functionality. The use of water-based sol–gel coatings and low-temperature plasma treatments, as well as solvent-free chemical vapor deposition, can be classified as green manufacturing because of the adoption of environmentally friendly deposition techniques. AI-assisted process optimization concerning coatings is projected to decline energy expenditure, material use, and emissions to aid the mitigation of the overall production process environmental footprint [28]. Surface engineering, within the context of the Fourth Industrial Revolution, is increasingly becoming a cyber-physical discipline. Materials, machines, and data comprise an integrated and interlinked ecosystem to surface engineering. The Internet of Things facilitates communication between smart surfaces and control systems, enabling real-time monitoring and control of complete production systems. AI-powered cloud-based analytics optimize self-sensing coated systems' smart lubrication systems, predict wear, and recommend process changes in real time. Intelligent systems controlling tribological processes will provide predictable operational reliability.

Moreover, they will facilitate improvements in global resources and energy efficiencies on a large scale, participate in the circular economy, and help shift towards globally sustaining technologies [29].

1.5 OPPORTUNITIES AND CHALLENGES

Advancements in AI, IoT, and cloud computing technologies provide the capability to create self-adjusting, data-responsive, smart adaptive surface engineering, and tribological systems for self-monitoring, predictive wear, and performance optimization. These systems are geared for friction loss, operational lifespan improvement, and increased efficiency in the manufacturing, transport, and energy sectors. However, the various opportunities along smart surface developments and their friction and wear engineering applications in modern industry are complemented with challenges on data holism, sensor fidelity, and system heterogeneous integration, as well as costs and the need for interdisciplinary efforts. This chapter answers research questions by explaining the drivers and challenges of friction and wear engineering, modern surface engineering, and smart surface developments in contemporary industrial applications [30]. Figure 1.6 shows the integration of artificial intelligence, Internet of Things, and cloud computing technologies to set up smart adaptive surface engineering systems with self-assessment, predictive wear monitoring, and performance adjustment for manufacturing, transport, and energy.

Integrating AI, IoT, and cloud technologies for automated friction and wear optimization is the most transformative of these opportunities. During the earlier passive phase of monitoring tribological system performance, the maintenance tribologist made periodic evaluations and adjustments. Now, the maintenance tribologist can leverage AI- and IoT-embedded technologies for continuously evaluating surface conditions in real time. The embedded sensors measure temperature, vibration, lubrication thickness, wear rate, and a good many other parameters, transmitting the data to AI-equipped cloud systems for abnormality, failure prediction, and prescription for corrective action. AI-based machine learning and deep learning systems can be historically trained to autonomously optimize intervals for lubrication, surface texture, and coating configuration. The reduction of maintenance downtime and costs, operational inefficiencies, and the improved reliability of the maintained system are

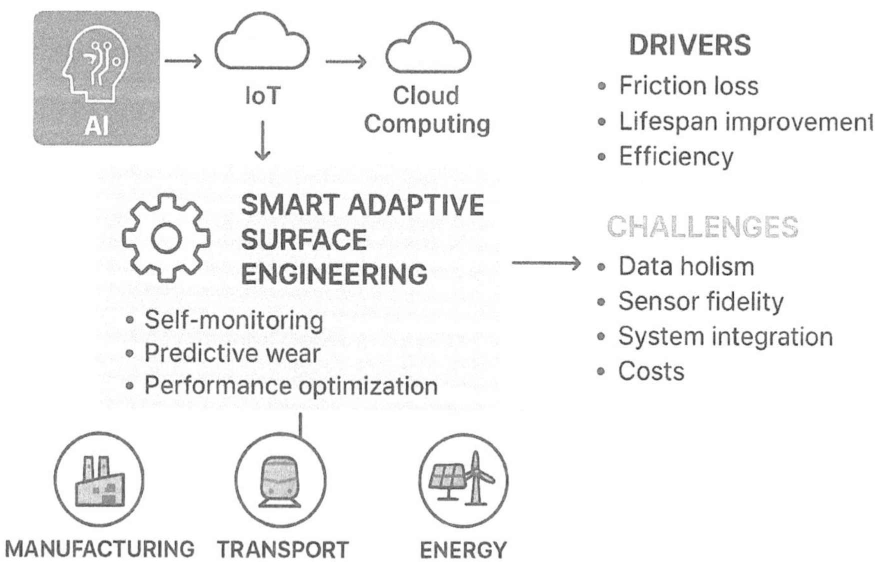

FIGURE 1.6 Integration of AI, IoT, and cloud computing in smart adaptive surface engineering.

direct outcomes of the proposed predictive and prescriptive systems. AI-tribology systems prove self-regulation by continuously monitoring and adjusting tribological parameters, such as lubrication and temperature control, to avoid premature failure in high-speed rotating machinery [31].

The emergence of self-reporting, self-healing, and adaptive surfaces demonstrates yet another paradigm shift within the sector. The forthcoming variants of functional coatings and surface materials will incorporate sensing and responsive capabilities within the surfaces themselves. For instance, self-reporting surfaces that include micro- or nanoscale sensors are able to sense and provide constant feedback on the progression of wear, corrosion, and fatigue. These self-reporting surfaces can connect to AI monitoring systems to interpret real-time data and provide alerts regarding maintenance or operational adjustments. Self-healing coatings that contain microcapsules of healing agents or polymer networks that contain covalent bonds of some reversible order can autonomously heal microcracks and surface damage without the need for human intervention. This improves the durability of components, and the sustainable aspect of manufacturing will be further supported by reductions in the frequency of component replacements. Finally, adaptive surfaces can change the frictional, thermal, or hydrophobic properties of the surface in response to the surroundings. They hold great promise in environmentally controlled systems for the aerospace industry, precision instruments, and biomedical implants. For instance, temperature-responsive coatings can change lubrication performance and frictional behavior, adapting to thermal load variations and fluctuating conditions in real time [32].

A new and promising area involves the application of digital twins for simulation-based design and predictive optimization of tribological systems. Digital twins refer to virtual representations of physical systems with real-time sensors. Combining real-time data and computational models allows engineers to estimate wear, friction heating, and surface degradation over time under several conditions, eliminating the need for physical experimentation. Such simulation-extensive designs are cheaper, and surface engineering predictive surfaces are becoming more accurate. In more advanced environments, the digital twin concept allows modeling not only individual surfaces but entire mechanical assemblies, thus rendering system-level frictional interaction analysis and surface engineering spatial design more efficient. The fusion of artificial intelligence with digital twins allows for adaptive learning, where new data alters the predictive model to accurately reflect real-world behavior [33].

Advanced surface engineering, through the development of low-friction coatings, self-lubricating composites, and wear-resistant nanostructures, leads to the direct reduction of energy use and the mitigation of greenhouse gas emissions. The use of AI for lubrication optimization is another way to limit environmental pollution, as it assesses the precise amount to be used and the timing to be used in the process. Then, smart coatings and reconfigurable surfaces assist in the reuse and refurbishment of system components, a core element of circular economies. Self-diagnosing and self-repairing components substantially extend service life and, therefore, reduce waste of materials. Likewise, energy-efficient manufacturing processes, which include microwave cladding, sol–gel deposition, and laser-assisted coating, help achieve high-performance surfaces with reduced energy input, making sustaining production goals more attainable in the aerospace, automotive, and renewable energy industries [34].

Implementing intelligent systems in tribology continues to face promising challenges. Multifaceted challenges in implementing intelligent systems in tribology are promising hurdles to overcome. Notable are challenges surrounding data standardization and safe communication between tribological sensors and digital systems. Closed-loop control in smart ecosystems necessitates data flow between real-time monitors and control systems. The absence of universal compatibility in sensor systems and communication protocols results in systems and software incompatibility. Furthermore, to maintain operational and safety integrity, no unauthorized access or control of system functionality should be allowed. Collateral operational loss or safety compromise can be caused by unauthorized access or control of operational data. Intrusion [25] and pervasive sensor-enabled systems make data and access control compromise easily actionable. Effective standardization of data communication and secure flow systems will make protection feasible in dense networking and control of data systems [35].

The complexity and price of smart sensing hardware significantly impede progress. Coatings and other structural components that contain embedded miniaturized sensors require specialized design and advanced integration methods, while the sensors themselves must survive extreme operational conditions such as high temperature, mechanical stress, and corrosive environments. These conditions drive design, maintenance, and operational costs. As sensors age and degrade, the reliability of predictive maintenance systems becomes compromised. Even though miniaturized sensors and other mechanisms embedded in smart hardware components surveyed within the last few months provide the necessary cost reductions, implementing these systems within the process remains economically impractical. As such, research within sensor systems must shift into designing cost-effective, energy-efficient, and structurally durable sensors that require little maintenance while operating in the extreme conditions found within most industrial systems. Designing, implementing, and maintaining intelligent tribological systems also require multidisciplinary expertise, which presents yet another problem. There is a gap within the education and training of surface engineers that spans the disciplines of chemistry, coating mechanics, data acquisition, AI modeling, and cloud integration, as well as within data science and the tribological sciences. Interdisciplinary training and collaboration are some of the key factors that slow the innovative and rapid adoption of new technologies. For the cultivation of hybrid skill sets needed for this field, the development of cross-disciplinary educational training, collaborative research, and partnerships between industry and academia are needed [36].

Incorporating AI and IoT in tribological systems also poses challenges with model validation and interpretability. Although machine learning models are capable of powerful predictions, many remain "black boxes," and users cannot discern the relationship between the inputs and outputs. This lack of interpretability can hinder the use of machine learning in mission-critical biomedical and aerospace applications, where the systems' reliability and traceability are of utmost importance. AI for X tribology (specialized tribology) needs to advance to provide the needed model explainability [37].

1.6 SUSTAINABLE SURFACE ENGINEERING AND TRIBOLOGY IN INDUSTRY 4.0

The globalization of sustainable practices and the expansion of modern manufacturing have increasingly integrated sustainability considerations in deciding the future of manufacturing processes, materials, and systems under the influencing factor of Industry 4.0. In this scope of sustainable surface engineering and sustainable tribology, focus on the sustainable development goals to curtail the adverse effects of friction, wear, and waste on the environment, materials, and energy (Figure 1.7). Tribology encompasses the loss of 23% of the total energy consumed worldwide, which exemplifies an increased focus on the development of smart, adaptable, and energy-efficient surface technologies. The integration of Tribology 4.0 is pivotal to balancing sustainability and industrial development [38].

In the context of Industry 4.0, the principles of sustainable surface engineering go beyond the strengthening of surface wear. The design of surfaces that are energy efficient, self-healing, recyclable, and resource minimizing throughout their life cycle is the next advancement in this field. The combination of artificial intelligence, the Internet of Things, and digital twins fosters the active monitoring, predictive maintenance, and life cycle optimization of tribological surfaces [40]. These

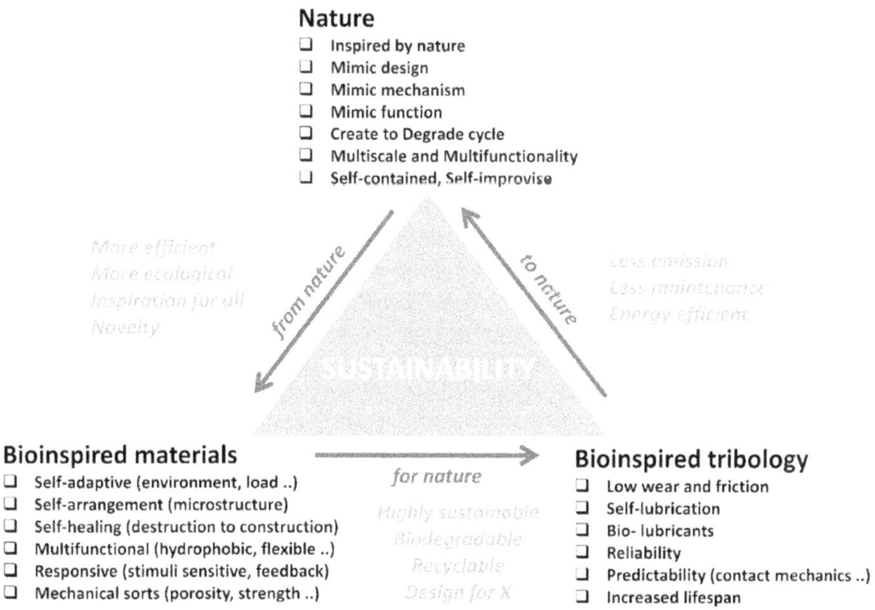

FIGURE 1.7 Sustainability triangle with nature, materials, and tribology as its corners [39].

predictive tools allow manufacturers to forecast surface wear and maintain surfaces in the period prior to a critical failure, optimizing parameters to eliminate excess. AI optimization of thermal spraying and laser cladding is similar to the control of thermal spray coatings where parameters such as thickness, microstructure, composition of the coating, and energy during production are all accounted for to minimize excess. IoT sensors embedded in the coating and connected to the cloud analyze friction, temperature, and load to control the coating, while machine learning monitors the processes to control and eliminate excess. An important focus on sustainability in tribology is the innovation of green technologies for coatings and lubricants. Unlike modern sustainable tribological techniques, which primarily use low energy sol–gel processing, cold spraying, and microwave cladding, traditional lubricants and coating processes utilize high-energy treatments and chemicals that are harmful to the environment, such as VOCs and other chemicals that are toxic and/or involve energy-intensive processes at high temperatures. For example, microwave-assisted surface processing that minimizes full energy treatments requires far less energy than traditional thermal treatment and more thermal energy is concentrated on the surface that needs to be treated [41, 42]. Sol–gel coatings' wear and corrosion resistant properties are far better than their toxic solvent counterparts and are far better as they use water-based precursors. Utilization of bio-based lubricants and solid lubricants has become instrumental in sustaining tribology. Natural oils, esters, and biopolymers are replacing petroleum lubricants due to biodegradability and non-toxicity, and solid lubricants such as graphene, molybdenum disulfide (MoS_2), and hexagonal boron nitride (h-BN) are increasingly used in coatings to form self-lubricating surfaces that can function with very little or no liquid lubricants. This minimizes the use of lubricants, decreases contamination, and makes waste disposal easier. The incorporation of self-healing and self-lubricating coatings, which can autonomously lubricate under stress or heal damaged areas, and self-replenishing lubricating coatings prolong the lifespan of components and reduce the frequency of replacements needed, significantly supporting the principles of a circular economy [43]. FGMs allow the transition focal point from surface hardness to core ductility, which fosters mechanical integrity while reducing the materials as compared to traditional multilayer systems. Reduction is very important in the construction of surface engineering as it promotes sustainability. Out of the various manufacturing processes in construction, flexible grading construction processes such as directed energy deposition and selective laser melting preserve and control the microstructural and compositional characteristics of the surface materials. These processes apply to construction as functionally graded materials integration and application in surface engineering reduce material waste and promote the sustainability of surface engineering. AM is more sustainable because of its near-net-shape fabrication, which minimizes material waste, allows for the recycling of raw powders, and reduces material usage. Hybrid methods combining AM with laser or plasma coatings are capable of resurfacing worn components to restore and prolong their life cycles, decreasing the demand for new materials [44, 45].

In tribology, the implementation of circular manufacturing models further enhances sustainability. In this case, materials and components are designed for disassemblability, remanufacturability, and recyclability, as opposed to disposal. There are smart coatings and surface treatments that offer the opportunity to replace surfaces functionally without full replacement. AI models help predict restorative surface treatments and strategically time interventions to maximize the value of resources throughout a component's service life. The closed-loop resource cycle minimizes newly mined materials, lessens the ecological footprint, and allows for the economy of the cycle of resource manufacturing to be sustainable. For sustainable surface engineering, the primary focus should be on the operational phase—especially minimizing energy consumption and reducing emissions generated during service. The power loss in the systems of the steam engine, the gas turbine, and industrial athletes can be eliminated by surface engineering and adaptive lubrication systems. For example, laser-textured surfaces can create microlubrication films that reduce power loss by decreasing frictional energy dissipation and function under dry or boundary lubrication conditions, drastically decreasing power loss. The energy-saving potential in the friction-reduction technologies is significant, particularly for the transport and manufacturing sectors of the economy and their associated global carbon footprint. To integrate the science of materials, digital technologies, and environmental engineering in pursuit of a truly sustainable paradigm of surface engineering in the context of Industry 4.0, system-level holonomic engineering is vital. Uniform methodologies need to be developed to address the environmental surface technologies holistically, inclusive of primary resource extraction and recycling. The environmental assessment of materials should be integrated into the design phase along with the Life cycle assessment (LCA) to facilitate sustainable material selection by gauging the energy cost of the material, potential emissions, and recyclability. Simultaneously, the rising use of eco-friendly coating technologies combined with green tribological materials in various industries depends on the existence of supportive regulations and environmental incentives in the relevant industries [46].

1.7 CONCLUSION

In modern industries, Tribology 4.0 offers innovative solutions for dealing with friction, wear, and lubrication. For the first time, friction modifications were made possible by digital technologies, smart and sensing materials, and real-time predictive and prescriptive analytics. Recent and powerful AI, machine learning, and neural network science enable friction surface engineering to advance to self-sensing, self-healing, and adaptive surfaces that frictionally respond and adjust to changing operational conditions. Progress in surface engineered friction and wear components includes laser surface texturing, additive manufacturing, and nanocomposite functional and bio-lubricated coatings. Sustaining technological advancement by overcoming eco-tribological challenges is evidenced by the development of eco-friendly coated surfaces and bio-lubricants. Together with new eco-tribological challenges, the new principles of green manufacturing and green engineering design straddle friction surfaces for performance and recycling, recovery, and reuse to low carbon and circular materials and components. Closed loop systems of controlled optimized materials and energy are sustained to reduce use. Systems embedding AI, digital twins, and IoT are self-sustaining, responsive, and adjustable to enable new controlled eco-tribological challenges. Numerous challenges still need to be addressed, including the creation of standardized data protocols, the full integration of miniaturized, robust sensor systems, the development of cyber secure IoT infrastructures, and the provision of interdisciplinary integration of materials science, data science, and mechanical engineering. The development of intelligent conforming and sustainable tribological solutions will depend on cross-sector collaboration involving academic, industrial, and government stakeholders. The marriage of Tribology 4.0 and sustainable surface engineering lays the foundation of advanced smart manufacturing, complementing one another to significantly increase the operational lifetimes of manufacturing components, reduce total energy waste, and minimize waste on the production floor. By constructing intelligent and adaptive surface technologies, the future of manufacturing will not only be operationally efficient but also ecologically sustainable.

REFERENCES

1. Lisiecki, A. "Tribology and surface engineering." *Coatings 9 no.* 10 (2019): 663.
2. Kumar, D., 2018. "*Surface engineering: Enhancing life of tribological components.*" Springer, India, Private.
3. Bala, A., Rashid, R.Z.J.A., Ismail, I., Oliva, D., Muhammad, N., Sait, S.M., Al-Utaibi, K.A., Amosa, T.I. and Memon, K.A. Artificial intelligence and edge computing for machine maintenance-review. *Artificial Intelligence Review*, 57, no. 5 (2024): 119.
4. Kano, Seisuke. "A short review: Tribology in machining to understand conventional and latest modeling methods with machine learning." *Machines* 13, no. 2 (2025): 81.
5. Walia, R. S., Qasim Murtaza, Shailesh Mani Pandey, and Ankit Tyagi, eds. *Surface Engineering: Methods and Applications.* CRC Press, 2022.
6. Kayode, J. F., S. L. Lawal, S. A. Afolalu, and E. R. Racheal. "Essentials and future progress of green tribology in industry 4.0." Proceedings of the 2024 International Conference on Science, Engineering and Business for Driving Sustainable Development Goals (SEB4SDG), IEEE, April 2, 2024, pp. 1–11.
7. Dobrzański. L. A., and A. Dobrzańska-Danikiewicz. "Applications of laser processing of materials in surface engineering in the industry 4.0 stage of the industrial revolution." *Materials Performance and Characterization* 8, no. 6 (June 1, 2019): 1091–1129.
8. Mohanraj, T., and R. Jegadeeshwaran. "Introduction to industry 4.0." In *Tribology of Polymer and Polymer Composites for Industry 4.0*, edited by H. Jena, J. K. Katiyar, and A. Patnaik, 113–127. Singapore: Springer, 2021.
9. Chakraborty, Koushik, Dhiraj Kapila, Sumit Kumar, Bhupati, Nazeer Shaik, and Akanksha Singh. "Intelligent machine learning based Internet of Things (IoT) resource allocation." *Engineering Proceedings* 59, no. 1 (2023): 73.
10. Al-Samarai, R. A., and Y. Al-Douri. *Metal Surface Engineering: Developed Tribology.* CRC Press, 2025, May 30.
11. Sellés Cantó, M. Á., and S. Sánchez-Caballero. *Book of Abstracts from the 10th International Conference on Tribology in Manufacturing Processes and Advanced Surface Engineering.* Trans Tech Publications Ltd, 2024.
12. Raja, Subramani, Ronit Rosario Leon, Nageswaren Rajeswari, Maher Ali Rusho, and Shankar Karthik Venkitaraman. "Tribological performance enhancement in FDM and SLA additive manufacturing: materials, mechanisms, surface engineering, and hybrid strategies—a holistic review." *Lubricants* 13, no. 7 (2025): 298.

13. Monikandan, V. V., K. Pratheesh, Carlo Santulli, Sonia Digra, P. V. Prasanth, and E. R. Dhas. "History of tribology: Assessing the prehistoric impacts, progress on industrial and scientific revolution eras, and contemporary interdisciplinary research trends." *Tribologia-Finnish Journal of Tribology* 41, no. 3–4 (2024): 4–15.
14. Dobrzański, L. A. "Role of materials design in maintenance engineering in the context of industry 4.0 idea." *Journal of Achievements in Materials and Manufacturing Engineering* 96, no. 1 (2019): 12–49.
15. Kano, Seisuke. "A short review: Tribology in machining to understand conventional and latest modeling methods with machine learning." *Machines* 13, no. 2 (2025): 81.
16. Ta, Thi-Na, Shin-Yuh Chern, and Jeng-Haur Horng. "Tribological behavior of ionic liquid with nanoparticles." *Materials* 14, no. 21 (2021): 6318.
17. Marian, Max, and Stephan Tremmel. "Current trends and applications of machine learning in tribology—A review." *Lubricants* 9, no. 9 (2021): 86.
18. Alam, M. S. "Industrial tribology in the past, present and future: A review." *Material Science & Engineering* 5, no. 3 (2021): 85–91.
19. Johns-Rahnejat, Patricia M., Ramin Rahmani, and Homer Rahnejat. "Current and future trends in tribological research." *Lubricants* 11, no. 9 (2023): 391.
20. Cheng, Jie, Shuo Gao, Jiahui Ma, Zhenxiang Huang, and Shirong Ge. "Electronics manufacturing tribology: Future opportunities and challenges." *Friction* 13, no. 12 (2024): 9441048.
21. Awotunde, Joseph Bamidele, Kamalakanta Muduli, and Biswajit Brahma. *Computational Intelligence for Industry 4.0 and 5.0 Applications.* Taylor & Francis Group, 2024.
22. Wu, Ze, Hang Bao, Youqiang Xing, and Lei Liu. "Tribological characteristics and advanced processing methods of textured surfaces: A review." *The International Journal of Advanced Manufacturing Technology* 114, no. 5 (2021): 1241–1277.
23. Dong, Jialin, Yang Liu, and Manuela Pacella. "Surface texturing and wettability modification by nanosecond pulse laser ablation of stainless steels." *Coatings* 14, no. 4 (2024): 467.
24. Jiao, Roger, Sesh Commuri, Jitesh Panchal, Jelena Milisavljevic-Syed, Janet K. Allen, Farrokh Mistree, and Dirk Schaefer. "Design engineering in the age of industry 4.0." *Journal of Mechanical Design* 143, no. 7 (2021): 070801.
25. Holmberg, Kenneth, and Allan Matthews. *Coatings Tribology: Properties, Mechanisms, Techniques and Applications in Surface Engineering*. Vol. 56. Elsevier, 2009.
26. Kumar, Kaushik, Divya Zindani, and J. Paulo Davim. *Industry 4.0: Developments Towards the Fourth Industrial Revolution.* Springer, 2019.
27. Mousavi, Yalda, Zahra Gharineiat, Armin Agha Karimi, Kevin McDougall, Adriana Rossi, and Sara Gonizzi Barsanti. "Digital twin technology in built environment: A review of applications, capabilities and challenges." *Smart Cities* 7, no. 5 (2024): 2594–2615.
28. Fox-Rabinovich, German, George C. Weatherly, and Anatoli Kovalev. "Tribology and the design of surface-engineered materials for cutting tool applications." In Rafael Colás & George E. Totten (Eds.), *Modeling and Simulation for Material Selection and Mechanical Design*, 297–372. CRC Press, 2003.
29. Richert, Maria, Marek Dudek, and Dariusz Sala. "Surface quality as a factor affecting the functionality of products manufactured with metal and 3D printing technologies." *Materials* 17, no. 21 (2024): 5371.
30. Ghosh, Sourish, Ricardo Knoblauch, Mohamed El Mansori, and Cosimi Corleto. "Towards AI driven surface roughness evaluation in manufacturing: A prospective study." *Journal of Intelligent Manufacturing* 36, no. 7 (2025): 4519–4548.

31. Yang, Yuhang, Zhiqiao Dong, Yuquan Meng, and Chenhui Shao. "Data-driven intelligent 3D surface measurement in smart manufacturing: Review and outlook." *Machines* 9, no. 1 (2021): 13.
32. Tripathi, Shivi, Ansheed Raheem, Madhusmita Dash, Prasoon Kumar, Ahmad Elsebahy, Harpreet Singh, Geetha Manivasagam, and Himansu Sekhar Nanda. "Surface engineering of orthopedic implants for better clinical adoption." *Journal of Materials Chemistry B* 12, no. 44 (2024): 11302–11335.
33. Wang, Jian. "Introductory chapter: Unveiling present and future horizons of surface engineering." In *Surface Engineering-Foundational Concepts, Techniques and Applications*. IntechOpen, 2025. DOI: 10.5772/intechopen.1010857
34. Shah, Saumya, Hemant Nautiyal, Gaurav Gugliani, Ashwani Kumar, Tanuj Namboodri, and Yogesh Kumar Singla, eds. *Sustainability in Smart Manufacturing: Trends, Scope, and Challenges*. CRC Press, 2024.
35. Eusufzai, Zobayer. "Impact of advanced lubrication management systems on equipment longevity and operational efficiency in smart manufacturing environments." *ASRC Procedia: Global Perspectives in Science and Scholarship* 1, no. 01 (2025): 618–653.
36. Marian, M., and S. Tremmel. "Current trends and applications of machine learning in tribology–a review. lubricants 2021, 9, 86." *Machine Learning in Tribology* 165 (2021): 86.
37. Ssekasamba, Hakim, Mike Tebyetekerwa, Cui Haodong, Andrew Balilonda, Shafik Ssegwanyi, Qiu Gao, and Tang Xiaoliang. "Review of plasma surface engineering technology toward sustainable textile materials." *Materials Circular Economy* 6, no. 1 (2024): 27.
38. Parmar, Hetal, Tayyab Khan, Fausto Tucci, Rehan Umer, and Pierpaolo Carlone. "Advanced robotics and additive manufacturing of composites: Towards a new era in Industry 4.0." *Materials and Manufacturing Processes* 37, no. 5 (2022): 483–517.
39. Kumar, Rahul, Mansoureh Rezapourian, Ramin Rahmani, Himanshu S. Maurya, Nikhil Kamboj, and Irina Hussainova. "Bioinspired and multifunctional tribological materials for sliding, erosive, machining, and energy-absorbing conditions: A review." *Biomimetics* 9, no. 4 (2024): 209.
40. Soori, Mohsen, Roza Dastres, Behrooz Arezoo, and Fooad Karimi Ghaleh Jough. "Intelligent robotic systems in industry 4.0: A review." *Journal of Advanced Manufacturing Science and Technology* 4, no. 3 (2024): 2024007.
41. Horovistiz, Ana, and J. Paulo Davim. "5 Digital tribology and the industry 4.0–a note." *Manufacturing in Digital Industries: Prospects for Industry 4.0* 6 (2020): 97.
42. Kumar, Ashwani, Yatika Gori, Nitesh Dutt, Yogesh Kumar Singla, and Ambrish Maurya, eds. *Advanced Computational Methods in Mechanical and Materials Engineering*. CRC Press, 2021.
43. Modad, Ossama Abou Ali, Jason Ryska, Abdallah Chehade, and Georges Ayoub. "Revolutionizing sheet metal stamping through industry 5.0 digital twins: A comprehensive review." *Journal of Intelligent Manufacturing* 36, no. 6 (2025): 3717–3739.
44. Martínez-Mireles, Josué Román, Jazmín Rodríguez-Flores, Marco Antonio García-Márquez, Arturo Austria-Cornejo, and Rafael Alfonso Figueroa-Diaz. "AI in smart manufacturing." In *Machine and Deep Learning Solutions for Achieving the Sustainable Development Goals*, 463–494. IGI Global Scientific Publishing, 2025.
45. Ambadekar, Prashant K., Sarita Ambadekar, C. M. Choudhari, Satish A. Patil, and S. H. Gawande. "Artificial intelligence and its relevance in mechanical engineering from Industry 4.0 perspective." *Australian Journal of Mechanical Engineering* 23, no. 1 (2025): 110–130.
46. Doharey, Alok Kumar, and Ravi Shankar Rai, "Industry 4.0 and digital manufacturing." *Machining and Tribology of Advanced Materials: From Coatings, Lubrications, Surface Treatments to Modeling and Simulation* 12 (2025): 57.

2 Transforming Tribological Systems through Digital Technologies, Digital Twins, and Innovations

Amrinder Mehta, Hitesh Vasudev, Jashanpreet Singh, and Brijesh Prasad

2.1 INTRODUCTION

The future of manufacturing is fast changing following the digital transformation and Industry 4.0 technologies. The secret behind successful digital manufacturing lies in the holistic nature of linking all the necessary technologies throughout an organization to the strategic partners rather than providing the traditional isolated implementations [1].

Tribology has been and remains one of the most topical areas in modern society as it is found in virtually every sphere of our life. The significance of friction, lubrication, and wear is also indicated by the large portion of current world energy consumption. The knowledge of tribology can open the door to new solutions to technical problems in the future. The foundation of all developments is millions of experiments and sophisticated computer simulations on various scales and in various fields of physics [2]. Because of the complexity of friction and wear, tribology is a very empirical science even though there have been numerous attempts to cstablish rules or laws of friction and wear. Nevertheless, the overload of friction, wear, and surface property information of various materials in tribological research presented a gateway to data analysis [3].

Recent developments in the application of techniques of deep learning (DL) or machine learning (ML) to enhance tribological properties of materials are very recent and achieved using artificial intelligence (AI). The focus of AI typically is on the construction and design of intelligent agents, anything that behaves in the most optimal manner in any circumstance. ML is a huge collection of data-intensive procedures and computational instruments to model and comprehend complicated datasets. Automatic patterns can be identified in the datasets through these methods,

DOI: 10.1201/9781003741152-2

thereby forming models that can be used to predict future data or other outcomes of interest in view of uncertainty [4].

Figure 2.1 presents the development of AI and its influence on tribology by means of combining learning algorithms and informatics strategies. It chronicles important advances of AI in 1956, machine learning in 1980, deep learning in 2006 and foundation models in 2021, including reinforcement learning, CNNs, transformers, and transfer learning. Multimodal information and few-shot learning are made possible by the fact that now large models of tribology can be used. At its base, the picture demonstrates how tribology evolved since computer-aided design, through neural network forecasting, to the current form of machine learning, and finally tribo-informatics, a discipline at the intersection of tribology with informatics to push forth innovative, data-directed solutions [5].

The existing conventional techniques to form these hybrid materials are largely based on a very ineffective, hit-and-miss mechanism. The recent data-driven methods of ML are becoming a new perspective for interpreting and developing more complicated processes within tribological research. The reason for this is that many more complex problems and datasets can be modeled with ease and at a low cost with greater adaptive capabilities to shifting conditions with the help of ML. These technologies enable virtual replication, simulation, and optimization of tribological processes, leading to improved system efficiency, reliability, and longevity across diverse industries [6].

Figure 2.2 depicts the connection and chain of command between AI, ML, and DL. It reveals that AI is the most extensive field as it includes all methods that allow

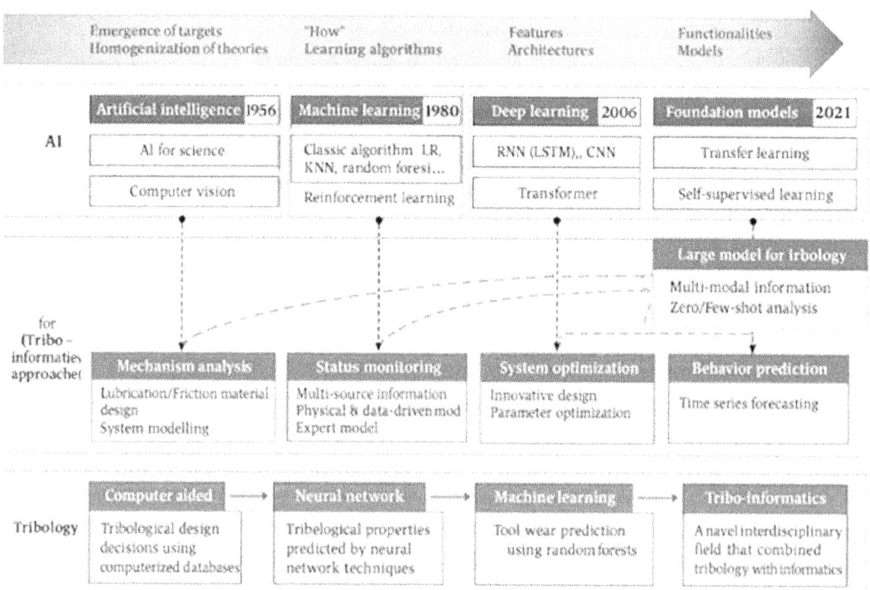

FIGURE 2.1 Development of AI in tribology [5].

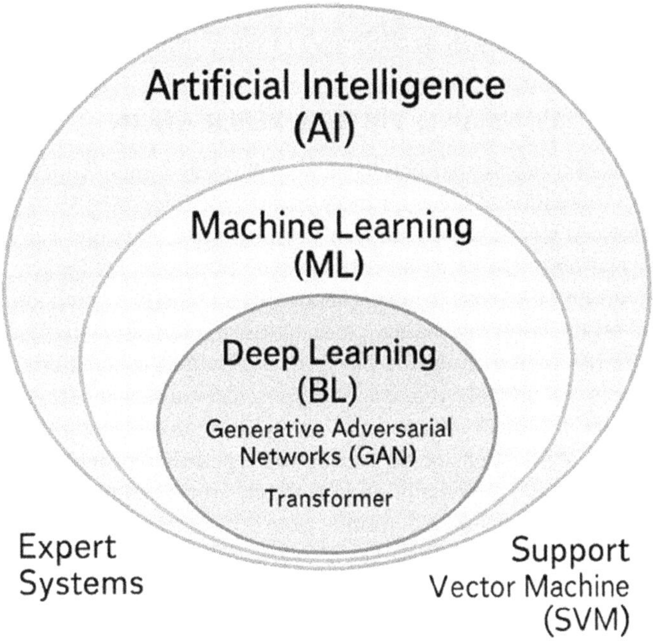

FIGURE 2.2 Diagram of the relationship between artificial intelligence, machine learning, and deep learning [7].

machines to work intelligently. To further elaborate, deep learning is a set of machine learning techniques that is specialized in the use of multilayered neural networks to estimate more complex patterns. The picture mentions examples of deep learning architecture, namely, generative adversarial networks (GANs) and transformers. Beyond the fundamental circles, it indicates that expert systems and support vector machines (SVMs) are subfields or methods related to AI and ML, respectively. The following nested diagram helps to understand the scope and relationships of these technologies in that deep learning lies within machine learning, which is a subset of the broader AI domain [7].

As mentioned earlier, large-scale studies were performed to determine the impact of input parameters on process efficiency and to understand the correlation between process input and output data. The global attention to the use of ML methods that can solve complex and nonlinear processes has inspired their application to optimize tribology processes [8].

Tribotronics makes use of so-called loss outputs like vibration, wear, and friction as feedback data to enhance efficiency and performance of the machinery. This involves active condition monitoring of a tribological system where autonomous and self-optimizing algorithms are used to implement corrective measures by actuation to obtain optimal performance [9].

Increased applications of sensors in tribological components and ML or AI will also enable the transition in the industrial tribological analytics. This review is a critical analysis of the way these digital innovations are changing tribological systems, the opportunities as well as challenges of implementing these innovations.

2.2 DIGITAL TWINS AND VIRTUAL REPLICATION

Digital twins (DTs) are virtual equivalents to physical things, which involve data collection and simulation of behavior in a dynamic and integrated setting. Regarding the unconventional gas fields, digital twins can help operators and engineers monitor, analyze, and optimize many aspects of field operations. DTs play a key role in realizing the vision of a smart factory. Classically, a DT is the virtual representation of a physical product, system, being, community, or even cities that are constantly updated with information about the physical version and its surrounding world. It fills the gap between virtual cyberspace and physical entities and, therefore, can be deemed the pillar of Industry 4.0 and the key to innovation of the future [10].

The numerous applications of DTs have been regularly reported in the literature, but the construction and application of DTs in real-world situations are not as notable. Moreover, a consensus on a definition of a DT does not exist, which gives rise to misunderstandings as to what its key characteristics are and how it differs from product models, simulation of the processes, and optimum use cases [11].

DTs are becoming beneficial to mechanical and aerospace systems by accelerating the development process, minimizing risk, preventing problems, and minimizing costs of sustainment. These advantages will not manifest themselves at scale unless a deliberate and formal process of conceiving, designing, developing, operating, and maintaining DTs is undertaken. A digital twin does not require being a beautiful virtual copy to add maximum value but rather should be imagined to be purposeful, with the choice of the fitness being determined by the required capabilities and the trade-offs between cost and value [12].

Figure 2.3 adopts the concept of an integrated digital twin and demonstrates how the different areas come together to facilitate the purposes of simulation, learning, and management. In the middle, there is a digital twin in the form of a computer chip, which reflects its connection to a physical system in having a physical counterpart. The diagram classifies supportive technologies and processes under the digital twin. The functions of simulation are represented in gray, and they consist of structural

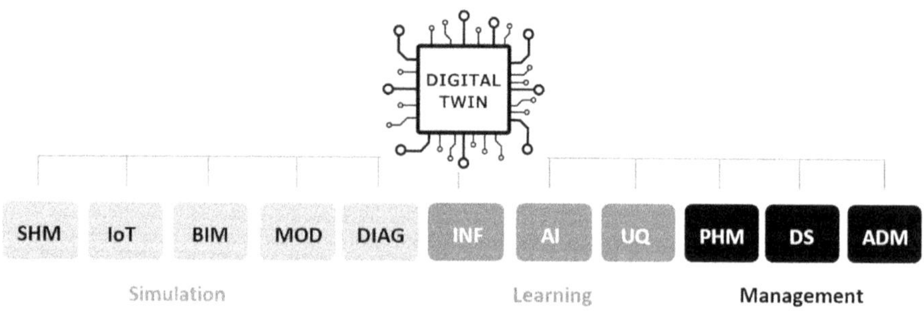

FIGURE 2.3 Digital twin conceptualization in terms of features and pillars [13].

health monitoring (SHM), Internet of Things (IoT), building information modeling (BIM), modeling (MOD), and diagnostics (DIAG). The learning processes (in blue) include information fusion (INF), artificial intelligence (AI), and uncertainty quantification (UQ). The management activities are highlighted in black, and they include prognostics and health management (PHM), decision support (DS), and asset management (ADM). This organization emphasizes the role of digital twins in bringing together data criteria of actual systems, utilizing clever learning algorithms, and authorizing well-informed choices in administration to streamline performance, dependability, and life cycle planning [13].

A DT is a system-of-systems that transcends the old-fashioned computer-based simulations and analysis. It is a simulation of the entire aspects, procedures, movements, and software of a physical structure into a computer version. The two systems (physical and digital) coexist, with all the inputs and operations being shared by means of real-time data communications and information transfer [14].

DT systems provide the opportunity to virtually recreate tribological systems, which makes it possible to do real-time simulation, prediction, and optimization of performance. These models combine sensor and manufacturing execution system data to develop dynamic models that replicate physical systems to support decision-making and optimization of processes. Digital twins in manufacturing allow management of errors, disassembly, and optimization online without interrupting a manufacturing process. Digital twins offer automated decision-making in structural engineering, energy systems, and integrity management. The availability of affordable sensors with high precision has enabled the real-time status of friction, wear, and lubrication to become feasible [15].

Smart sensors that may be meant to operate with IoT and cloud solutions allow acquiring constant data and monitoring conditions. Generated by its users, the prediction is that it will evolve to an intelligent platform. This will see work shift to the virtual world where there would be significant effects on efficiency and effectiveness.

2.3 AI, MACHINE LEARNING, AND TRIBO-INFORMATICS

Wear behavior fabrication and assessment is time and cost limited. Based on this, machine learning methods have become an important instrument to forecast the mechanical behavior, wear, and tribological behavior of various materials. The use of AI and ML has become a potent instrument in the study concerning tribology and has provided new prospects to improve the insights of wear behavior and forecast tribological characteristics [16].

Tribology has also not been left out in using ML and AI methods, which may assist in sorting out the intricacies of patterns and trends in the various interacting features and processes. It is concluded that the application of artificial intelligence and machine learning algorithms enhances the accuracy of the tribological system diagnostics significantly, and enables predicting its service life and making the operating parameters of tribological systems and machine mechanisms optimal [17].

Figure 2.4 defines the three major groups of machine learning, namely, supervised learning, unsupervised learning, and reinforcement learning. Supervised

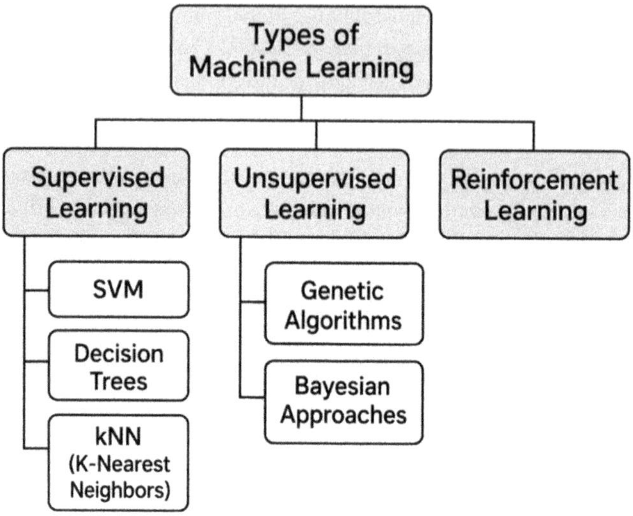

FIGURE 2.4 Structuring of ML techniques and algorithms [18].

learning is described as made up of algorithms that use labeled data to learn how to do things and includes SVM, decision trees, k-nearest neighbors (kNN), and neural networks. Unsupervised learning, in contrast, refers to the extraction of patterns or structures of datasets that do not contain explicit data, and encompasses genetic algorithms and Bayesian methods. The third branch is reinforcement learning, which describes algorithms that respond to the environment in such a way that they learn the best actions based on rewards. This flow chart-type diagram enables us to better understand the way to group different machine learning methods under these main categories and provides a general overview of the various approaches and models that data scientists and engineers may apply based on the type of data they have and the type of issues they are trying to address. The concise categorization will help to grasp the differences and similarities between the learning styles very fast [18].

Physics-informed machine learning (PIML) is a new method that combines physical laws with data-driven machine learning models. PIML is becoming an influential tool in many scientific and engineering disciplines by harnessing the opportunities presented by both classical machine learning and the knowledge of physical sciences. The major benefit of PIML is that it enhances the generalizability of the model and makes the predictions remain consistent with the laws of physics. This method prevents the weakness of data-driven methods, which can be large and cannot be generalized beyond the training domain [19].

Figure 2.5 describes the phenomenon of physics-informed machine learning that uses the combination of neural networks and physical laws to incorporate partial differential equations. On the left, the neural network receives the input variables, which are the spatial coordinates and time as inputs, and gives out outputs that are associated with the terms in the equations, such as the initial and the boundary conditions.

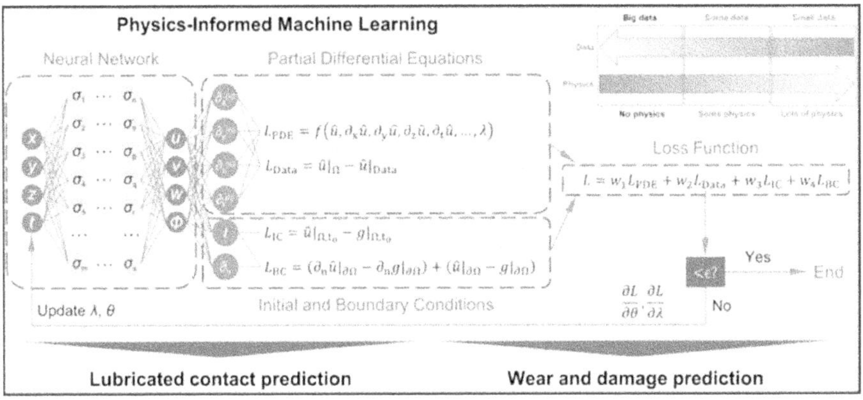

FIGURE 2.5 Physics-informed machine learning [20].

They are mathematically expressed as loss functions that are expressive of the various physical constraints and data accuracy, and weighted and added together to train the model. The right-hand side illustrates an iterative scheme: the model predictions are the ones against which someone evaluates the accuracy against a threshold, and in case this is not achieved, the neural network parameters are updated by the gradients of the loss function. Using physics, the method can be used to make strong predictions of lubricated contacts and wear or damage even in cases where data is not available [20].

Physics-informed neural networks (PINNs) are a trendy PIML approach. As an illustration, PINNs have already been used in fluid mechanics and proved to improve or substitute high-fidelity numerical models of turbulent flows [21]. Tribology is becoming revolutionized with AI and ML making it possible to analyze multiscale and multifaceted data to predict friction, wear, and lubrication performance. Tribo-informatics, which is a combination of tribology and informatics, helps in the maintenance of efficient resource usage and optimization of the systems [22].

Figure 2.6 explains a numerical proximity method of searching for the maximization of thermal conductance in substances via atomistic Green's functions analysis and Bayesian optimization. Two regions (A and B) with the atomic structures are shown on the left with particular emphasis on the interfacial region where heat transfer exists between the right lead and the left lead. Bayesian optimization has been used on the right. Each time, a randomly chosen set of candidate structures is created that consists of silicon (Si) and germanium (Ge) atoms. Consequently, the approach converts to an ideal atomic structure in the interfacial region to enhance thermal conductance functionality [23].

Digital technologies support the design and optimization of innovative tribological materials, such as composites, coatings, and lubricants. Predicting the performance of materials, optimizing process parameters, and faster identification of new materials with improved tribological characteristics are performed by means of machine learning models. The technologies of additive manufacturing and 3D

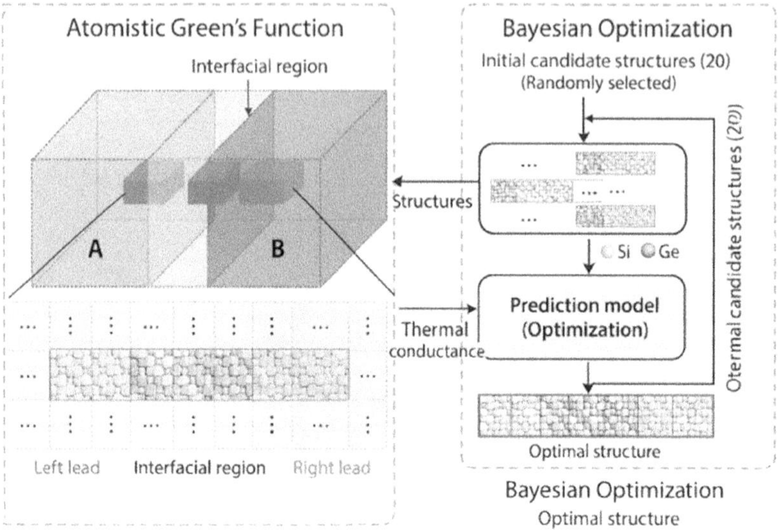

FIGURE 2.6 Schematic of the materials informatics approach integrating the atomistic Green's function (AGF) with Bayesian optimization [23].

printing, which follow digital models, can be used to optimally and quickly prototype tribological components [24].

2.4 DISCUSSION

Friction and wear manifestation in the real world, specifically in machinery, is primarily because of the presence of motion between surfaces that are touching each other. Therefore, the dynamics of components the surfaces belong to are a significant concern in the tribological state of the contacts of the surfaces [25].

Tribology is a significant aspect of heavy industry since it involves the use of an abrasive surfaces and rollers and is a highly interdisciplinary subject of study that intersects both theoretical and applied sciences through the integration of a wide range of subjects such as chemistry, mathematics, physics, materials science, and engineering. It affects everyday life but is quite unknown to the general public [26].

The integration of digital technologies into tribological systems is driving a paradigm shift from empirical, reactive maintenance to predictive, data-driven optimization. Digital twins and smart sensors provide unprecedented visibility into system behavior, enabling real-time monitoring and proactive interventions that reduce downtime and improve reliability [27].

It is possible to utilize coating technologies in different fields of engineering for enhanced surface and structural analyses. Advanced engineering analyses of surface and structural characteristics can be enhanced with characterizations of surface microtopographies and detailed assessments of mechanical properties involving tensile, compressive, and flexural strengths. Moreover, coatings can enhance the

TABLE 2.1
Approaches to Address Current Coating Industry Applications

Digital Technology	Application Area	Key Results	Reference
Cloud FEA, digital sensors	Lubricant evaluation in metal forming	Data-centric approach enables precise lubricant performance evaluation	[46]
AI, ML, tribo-informatics	System monitoring, material design	AI enhances reliability prediction and intelligent material design	[47]
Smart sensors, tribotronics	Real-time monitoring	Sensors enable condition monitoring and integration with Industry 4.0	[48]
Tribo-informatics, ML	System optimization	Informatics methods optimize materials, lubricants, and working conditions	[49]
Tribotronics, IIoT	Surface engineering	Smart surfaces enable in situ measurement and predictive maintenance	[50]

performance of solar energy systems by modifying their attributes. Furthermore, coatings can facilitate the numerical simulation of fluid flows through complex conduits like elbows and channel transitions in pipes. If the designs are for high-precision engineered artifacts, for example, those that are ergonomically designed such as seat systems, multifunctional coatings can improve the load distribution and enhance prediction of structural response and durability of the artifact, thus supporting further design innovation, optimization, and multiphysics simulation [28–45].

As shown in Table 2.1, digital technologies such as AI, ML, cloud finite element analysis (FEA), tribo-informatics, and smart sensors are applied in system monitoring, lubricant evaluation, optimization, and surface engineering.

The methods of AI and ML are especially useful to address the complexity and high dimensionality of tribological data and help in the design of improved materials and optimal work of the system. There is continuous concern about data quality, integration, and security, particularly in a large-scale industrial environment, whereas the thermal spray technique coating prevents the surface from corrosion and improves the hardness [51, 52]. Moreover, the use of such technologies is inefficient in all industries, and the obstacles comprise costs, knowledge, and system legacy.

2.5 CONCLUSION

The computerization of tribological systems is a revolutionary period in friction, wear, and lubrication studies and practices within any industry. Innovations such as digital twins, artificial intelligence, machine learning, smart sensors, and tribo-informatics can today be used to conduct comprehensive real-time monitoring, predictive maintenance, and smart optimization of materials and machinery. Digital twins are dynamic virtual representations of physical systems, which are beneficial in decision-making, simulation, and optimization of processes online. The complex tribological information sets can be processed by AI and machine learning algorithms; complex patterns are recognized, material behaviors are predicted, and the best operating parameters are proposed, which increases the reliability and efficiency of the system. With the introduction of smart sensors and IoT, ongoing data collection and proactive management of conditions are guaranteed, which simplifies the process of proactive intervention and increases the connectivity within the frames of Industry 4.0. Tribo-informatics is a bridge between tribology and informatics that enhances the optimization of the system and resource efficiency. Although this is good news, there are still issues of data quality, integrating the system, cybersecurity, and cost of implementation. The acceptance and ubiquity of digital technologies in tribology require standardization of the industry and sharing of knowledge. Finally, digital transformation in tribology opens avenues to smarter, more resilient, and sustainable engineering solutions.

REFERENCES

1. Yang, X., Liu, H., Dhawan, S., Politis, D.J., Zhang, J., Dini, D., Hu, L., Gharbi, M.M. and Wang, L., 2022. Digitally-enhanced lubricant evaluation scheme for hot stamping applications. *Nature Communications*, 13(1), p.5748.
2. Nosonovsky, M. and Bhushan, B., 2010. Green tribology: principles, research areas and challenges. *Philosophical Transactions of the Royal Society A: Mathematical, Physical and Engineering Sciences*, 368(1929), pp.4677–4694.
3. Hasan, M.S., Kordijazi, A., Rohatgi, P.K. and Nosonovsky, M., 2021. Triboinformatic modeling of dry friction and wear of aluminium base alloys using machine learning algorithms. *Tribology International*, 161, p.107065.
4. Biswas, S. and Satapathy, A., 2010. A study on tribological behaviour of alumina-filled glass–epoxy composites using Taguchi experimental design. *Tribology Transactions*, 53(4), pp.520–532.
5. Yin, N., Yang, P., Liu, S., Pan, S. and Zhang, Z., 2024. AI for tribology: Present and future. *Friction*, 12(6), pp.1060–1097.
6. Sose, A.T., Joshi, S.Y., Kunche, L.K., Wang, F. and Deshmukh, S.A., 2023. A review of recent advances and applications of machine learning in tribology. *Physical Chemistry Chemical Physics*, 25(6), pp.4408–4443.
7. Si, L., Niu, R., Dang, C., Bao, X., Zhuang, Y. and Zhu, W., 2024. Advances in artificial intelligence for artificial metamaterials. *APL Materials*, 12(12), p.120602.
8. Paturi, U.M.R., Palakurthy, S.T. and Reddy, N.S., 2023. The role of machine learning in tribology: a systematic review. *Archives of Computational Methods in Engineering*, 30(2), pp.1345–1397.
9. Liskiewicz, T., Sherrington, I., Khan, T. and Liu, Y., 2023. Advances in sensing for real-time monitoring of tribological parameters. *Tribology International*, 189, p.108965.
10. Jiang, Y., Yin, S., Li, K., Luo, H. and Kaynak, O., 2021. Industrial applications of digital twins. *Philosophical Transactions of the Royal Society A*, 379(2207), p.20200360.

11. Hyre, A., Harris, G., Osho, J., Pantelidakis, M., Mykoniatis, K. and Liu, J., 2022. Digital twins: representation, replication, reality, and relational (4Rs). *Manufacturing Letters*, 31, pp.20–23.
12. Ferrari, A. and Willcox, K., 2024. Digital twins in mechanical and aerospace engineering. *Nature Computational Science*, 4(3), pp.178–183.
13. Chiachío, M., Megía, M., Chiachío, J., Fernandez, J. and Jalón, M.L., 2022. Structural digital twin framework: formulation and technology integration. *Automation in Construction*, 140, p.104333.
14. Mihai, S., Yaqoob, M., Hung, D.V., Davis, W., Towakel, P., Raza, M., Karamanoglu, M., Barn, B., Shetve, D., Prasad, R.V., Venkataraman, H., Trestian, R. and Nguyen, H.X., 2022. Digital twins: a survey on enabling technologies, challenges, trends and future prospects. *IEEE Communications Surveys & Tutorials*, 24(4), pp.2255–2291.
15. Židek, K., Piteľ, J., Adámek, M., Lazorík, P. and Hošovský, A., 2020. Digital twin of experimental smart manufacturing assembly system for Industry 4.0 concept. *Sustainability*, 12(9), p.3658.
16. Pasha, M.B., Rao, R.N., Ismail, S., Gupta, M. and Prasad, P.S., 2024. Tribo-informatics approach to predict wear and friction coefficient of Mg/Si_3N_4 composites using machine learning techniques. *Tribology International*, 196, p.109696.
17. Aulin, V.V., Kovalov, S.G., Hrynkiv, A.V., Kovalov, Y.G., Holovaty, A.O., Kuzik, O.A. and Slon, V.V., 2025. Enhancing tribological system performance through intelligent data analysis and predictive modeling: A review. *Problems of Tribology*, 30(3/117), pp.49–61.
18. Wuest, T., Weimer, D., Irgens, C. and Thoben, K.-D., 2016. Machine learning in manufacturing: advantages, challenges, and applications. *Production & Manufacturing Research*, 4(1), pp.23–45.
19. Pateras, J., Rana, P. and Ghosh, P., 2023. A taxonomic survey of physics-informed machine learning. *Applied Sciences*, 13(12), p.6892.
20. Marian, M. and Tremmel, S., 2023. Physics-informed machine learning—an emerging trend in tribology. *Lubricants*, 11(11), p.463.
21. Yin, N., Xing, Z., He, K. and Zhang, Z., 2023. Tribo-informatics approaches in tribology research: A review. *Friction*, 11(1), pp.1–22.
22. Paturi, U.M.R., Palakurthy, S.T. and Reddy, N.S., 2023. The role of machine learning in tribology: A systematic review. *Archives of Computational Methods in Engineering*, 30(2), pp.1345–1397.
23. Ju, S., Shiga, T., Feng, L., Hou, Z., Tsuda, K. and Shiomi, J., 2017. Designing nanostructures for phonon transport via Bayesian optimization. *Physical Review X*, 7(2), p.021024.
24. Ganeshkumar, S., Deepika, T. and Haldorai, A., 2022. A supervised machine learning model for tool condition monitoring in smart manufacturing. *Defence Science Journal*, 72(5), pp.712–720.
25. Johns-Rahnejat, P.M., Rahmani, R. and Rahnejat, H., 2023. Current and future trends in tribological research. *Lubricants*, 11(9), p.391.
26. Adeolu, A.A., Kayode, J.F., Lawal, S.L. and Afolalu, S.A., 2024. Essentials and future progress of green tribology in Industry 4.0. In: *Proceedings of the 2024 International Conference on Science, Engineering and Business for Driving Sustainable Development Goals (SEB4SDG)*. pp.1–11.
27. Pai, R., Chattopadhyay, G. and Karmakar, G., 2022. Maintenance and asset management practices of industrial assets: importance of tribological practices and digital tools. *International Journal of Process Management and Benchmarking*, 13(2), pp.233–256.

28. Verma, P.K., Vasudev, H. and Kumar, V., 2025. Dental implant materials and designs and failure during implant placement and loading: A review. In: *Advanced Manufacturing Processes.* CRC Press, pp.138–150.
29. Verma, P.K., Vasudev, H. and Kumar, V., 2025. Bioactivity and corrosion performance of low-velocity oxy-fuel-sprayed hydroxyapatite-based coatings on a surgical-grade stainless steel. *Transactions of the Indian Institute of Metals*, 78(8), p.185.
30. Verma, P.K., Vasudev, H. and Kumar, V., 2025. Effect of electro-corrosion behavior of different grades of stainless steel used in bio-implants: A review. In: *Advanced Manufacturing Processes.* CRC Press, pp.269–283.
31. Verma, P.K. and Vasudev, H., 2025. Importance of hydroxyapatite coatings in improving new bone osteogenesis: A review. In: *Advanced Manufacturing Processes.* CRC Press, pp.284–299.
32. Dutta, V., Thakur, L., Singh, B. and Vasudev, H., 2022. A study of erosion–corrosion behaviour of friction stir-processed chromium-reinforced NiAl bronze composite. *Materials*, 15(15), p.5401.
33. Singh, M., Vasudev, H. and Kumar, R., 2021. Corrosion and tribological behaviour of BN thin films deposited using magnetron sputtering. *International Journal of Surface Engineering and Interdisciplinary Materials Science (IJSEIMS)*, 9(2), pp.24–39.
34. Sharanabasava, H., Raviprakash, M., Prasad, C.D., Ramesh, M.R., Phanibhushana, M.V., Vasudev, H. and Kumar, S., 2023. Microstructure, mechanical and wear properties of SiC and Mo reinforced NiCr microwave cladding. *Advances in Materials and Processing Technologies*, 10, pp.3620–3633.
35. Singh, M., Vasudev, H. and Singh, M., 2022. Surface protection of SS-316L with boron nitride based thin films using radio frequency magnetron sputtering technique. *Journal of Electrochemical Science and Engineering*, 12(5), pp.851–863.
36. Prashar, G., Vasudev, H., Thakur, L. and Bansal, A., 2023. Performance of thermally sprayed hydroxyapatite coatings for biomedical implants: A comprehensive review. *Surface Review and Letters*, 30(01), p.2241001.
37. Vasudev, H., 2020. Wear characteristics of Ni-WC powder deposited by using a microwave route on mild steel: Microwave cladding of Ni-WC. *International Journal of Surface Engineering and Interdisciplinary Materials Science (IJSEIMS)*, 8(1), pp.44–54.
38. Parkash, J., Saggu, H.S. and Vasudev, H., 2022. A short review on the performance of high velocity oxy-fuel coatings in boiler steel applications. *Materials Today: Proceedings*, 50, pp.1442–1446.
39. Singh, P., Vasudev, H. and Bansal, A., 2024. Effect of post-heat treatment on the microstructural, mechanical, and bioactivity behavior of the microwave-assisted alumina-reinforced hydroxyapatite cladding. *Proceedings of the Institution of Mechanical Engineers, Part E: Journal of Process Mechanical Engineering*, 238(3), pp.1060–1071.
40. Verma, P.K., Vasudev, H., Singh, G. and Singh, S., 2024. A review on the microstructural and biomedical properties of implants manufactured using additive manufacturing. *Journal of The Institution of Engineers (India): Series D*, 106, pp.1–15.
41. Singh, S.P., Verma, P.K., Singh, S., Kapoor, M. and Balu, R., 2024. Microstructural characterization and electrochemical behaviour of HVOF sprayed WC-Co-Cr coatings on 316L boiler steel stainless steel. *Results in Surfaces and Interfaces*, 16, p.100248.
42. Verma, P.K., Minhas, A.S., Singh, P. and Kumar, S., 2024, February. Slurry erosion behaviour of thermal sprayed Al2O3 and Cr2O3 coatings for turbine steels. *AIP Conference Proceedings*, 2986(1), p.020012.
43. Verma, P.K., Kumar, V. and Vasudev, H., 2024. Bioactivity and corrosion analysis of thermally sprayed hydroxyapatite based coatings. *Journal of Electrochemical Science and Engineering*, 14(4), pp.419–439.

44. Verma, P.K., Vasudev, H. and Kumar, V., 2025. Biocompatibility and corrosion performance of HA-based LVOF sprayed and post-annealed coatings on SS 254 surgical-grade bio-implant material. *Ceramics International*, 51(10), pp.13207–13225.
45. Verma, P.K., Vasudev, H. and Singh, G., 2025. Improving the corrosion performance of ss 304 stainless steel using HVOF-sprayed WC-Co-Cr coatings in chloride environments. *Journal of Bio-and Tribo-Corrosion*, 11(4), pp.1–13.
46. Liu, H., Yang, X., Politis, D.J., Shi, H. and Wang, L., 2024. Evaluation framework of digital characteristics (DC)-enhanced lubricant: Consideration of essential geometric features for hot-stamped components. *Journal of Manufacturing Systems*, 75, pp.150–162.
47. Ghritlahre, H.K. and Prasad, R.K., 2018. Application of ANN technique to predict the performance of solar collector systems – A review. *Renewable and Sustainable Energy Reviews*, 84, pp.75–88.
48. Hassija, V., Chamola, V., Mahapatra, A., Singal, A., Goel, D., Huang, K., Scardapane, S., Spinelli, I., Mahmud, M. and Hussain, A., 2024. Interpreting black-box models: A review on explainable artificial intelligence. *Cognitive Computation*, 16(1), pp.45–74.
49. Jani, D.B., Mishra, M. and Sahoo, P.K., 2017. Application of artificial neural network for predicting performance of solid desiccant cooling systems – A review. *Renewable and Sustainable Energy Reviews*, 80, pp.352–366.
50. Jatavallabhula, J.K., Gaonnwe, T., Nginda, S. and Veeredhi, V.R., 2025. Development and assessment of machine learning models for predicting fatigue response in AA2024. *Materials Research Express*, 12(1), p.016505.
51. Kovalev, A., Tian, Y. and Meng, Y., 2024. Prediction of ball-on-plate friction and wear by ANN with data-driven optimization. *Friction*, 12(6), pp.1235–1249.
52. Verma, P.K., Vasudev, H. and Kumar, V., 2025. Hybrid alumina-reinforced hydroxyapatite coatings deposited by using low-velocity oxy-fuel: Electrochemical corrosion and bioactivity assessment for biomedical applications. *Surface and Coatings Technology*, 519, p.132921.

3 Advanced Smart Coatings and Innovative Surface Treatment Technologies

Praveen Kumar Verma, Hitesh Vasudev, and Jürgen Markus Lackner

3.1 INTRODUCTION

Within contemporary manufacturing, the functionality and strength of surfaces determine component performance and longevity. More than 70% of the mechanical failures in the industrial business are surface-initiated failures due to friction, wear, and corrosion. Although traditional surface modifications such as coatings, and electroplated, anodized, and chemically converted surfaces offer some performance benefits, the modifications are largely inflexible and, over time, offer poor performance under variable conditions. Existing systems do not offer the needed performance trade-offs. The concept of smart coatings as surface modifications represents the first reasonable step in the addition of balanced integration of functionality and operational parameters for coatings [1]. The integration of nanotechnology and digital manufacturing offers the first meaningful opportunity to drive self-adjusting systems and sensors to offer self-regulating coatings as the foundational layer within the interventional smart materials of the Fourth Industrial Revolution. The systems offer self-repair and self-diagnosing capacities and real-time monitoring. This self-monitoring and self-adjusting capability predicts steps to reduce unplanned downtimes and component life cycle.

At the intersection of various engineering disciplines and research involving coated surfaces are the fundamentals of tribology: friction, wear, and lubrication [2]. The aerospace, automotive, energy, and specific biomedical engineering fields focus on developing surface coatings able to withstand extreme loads, high temperatures, and aggressive chemicals. State-of-the-art coatings protect surfaces, primarily at high temperatures, during sliding and boundary lubricated conditions while minimizing energy losses due to friction, thereby decreasing the 23% global energy consumption estimated for friction losses. Advanced coatings also align with the goals

 DOI: 10.1201/9781003741152-3

of sustainability, reduction of material waste, and less frequent repairs and maintenance. The ongoing Industry 4.0 transformation impacts the design and control of tribological systems. Coating techniques are increasingly optimized using concepts like digital twins and closed-loop control systems, while cyber-physical systems (CPS), the Internet of Things (IoT), and artificial intelligence (AI) provide the architecture for real-time design. Predictive machine learning (ML) models using large experimental and design datasets along with deposition parameters streamline the experimental process for coatings and predictive friction coefficients. The advance significantly accelerates the process of materials research and considerably reduces costs [3].

The performance of functional coatings has changed over time. Coatings now provide protective functions, hydrophobic properties, conductivity, and thermal stability. The development of coatings at the nanoscale, particularly those made of titanium nitride (TiN), diamond-like carbon (DLC), and graphene-reinforced composites, increased hardness and decreased friction. The degradation of other coating systems requires frequent external maintenance. Smart coatings, which are self-healing, self-lubricating, and stimuli-responsive, lose this detriment. Microencapsulation polymer coatings, for instance, are able to release and heal damaged coatings as well as stop corrosion. Phase-change lubricants change viscosity depending on load, temperature, and other preset conditions that contribute to passive maintenance of performance. Automated systems and contemporary robotics demand such properties. Control of the microbalance, chemistry, and morphology of a microstructure is a game changer for surface engineering. The ductility of a nanocomposite coating is a plus, as it compensates for the synergy of metallic, ceramic, and polymeric coatings to improve toughness and wear resistance. Incorporating nanoparticles—like SiC, Al_2O_3, ZrO_2, graphene, or carbon nanotubes—improves load-bearing capacity and lowers friction coefficients [4, 5]. Furthermore, nanostructured surfaces can be designed with hierarchical topographies that replicate natural systems (like lotus leaves or gecko feet) to gain self-cleaning or anti-adhesive properties. The recent development of atomic layer deposition (ALD) and pulsed-laser deposition (PLD) techniques makes it possible to produce ultra-thin films with atomic-level precision and conformality. These techniques are useful for achieving uniform coating and minimal defects on complex geometries of miniaturized or additive-manufactured parts [6]. Figure 3.1 provides a description of methods employed to modify the surface of metals according to the approach used, namely, mechanical, chemical, electrochemical, and thermal, and a brief overview of their contributions to the improvement of wear, corrosion, and tribological properties.

Various stimuli-responsive materials, including shape-memory alloys, electrochromic polymers, and magnetorheological composites, have begun to be incorporated into tribological systems for dynamic adaptability. Magnetorheological coatings, for instance, can modify their viscosity under the influence of a magnetic field, thus providing adjustable damping in precision actuators or systems designed for the control of vibrations. The incorporation of digital technologies across the entire coating life cycle from the design of processes to the assessment of performance provides the foundation for next-generation surface technologies. Digital

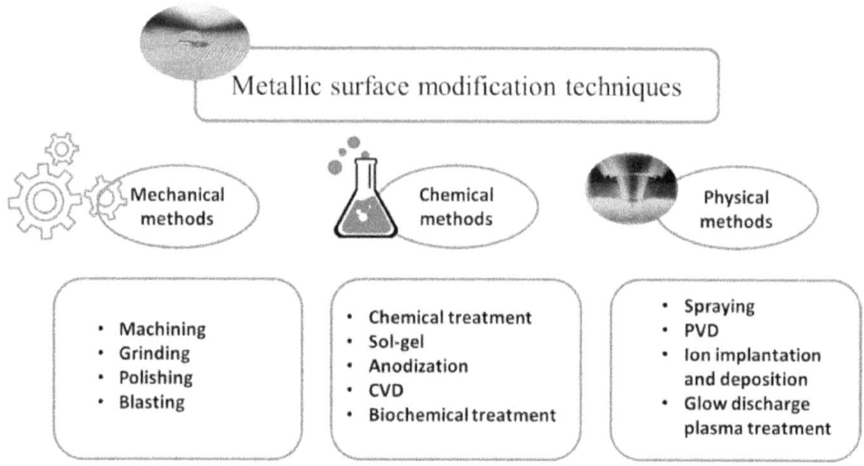

FIGURE 3.1 Classification of metallic surface modification techniques [7].

twins, virtual models of actual coating systems, have the capacity to project thermal profiles, stress distributions, and microstructural alterations throughout the deposition process. When coupled with real-time sensor data, such models facilitate adaptive process control to maintain consistent coating thickness and address quality issues [8].

AI and ML techniques have gained importance for analyzing large experimental datasets produced during the development of coatings. For example, supervised learning techniques have been employed to forecast tribological characteristics from attributes, such as spray distance, plasma power, or particle velocity. Currently, the use of reinforcement learning for autonomous optimization of process parameters in high-throughput coating systems is in the exploratory phase. This forms part of the complete automation envisioned by Industry 4.0: self-optimizing systems, predictive maintenance, and cyber-physical systems. These innovations highlight the significance of surface innovation as the link connecting material science to industrial value addition. With the advent of smart factories and automated inspection systems, the next generation of maintenance systems will integrate coatings that can relay their status to a central system [9]. Figure 3.2 is a schematic illustration that shows the sliding contact mechanism between two bodies, including the frictional forces, wear track, and contact interface.

This review consolidates recent scientific advancements and new technological applications in the area of hybrid smart coatings and surface treatments primarily focusing on the integration of Industry 4.0 and the advancements in the tribological coatings and surface treatments. The fundamentals of tribology and surface engineering are introduced in Section 3.2; the progression of coating and surface engineering technologies is outlined in Section 3.3; nanocoatings, self-healing coatings, and multifunctional composites are covered in-depth across Sections 3.4 to 3.7; and Sections 3.8–3.11 focus on the intersections of digitalization and sustainability, along

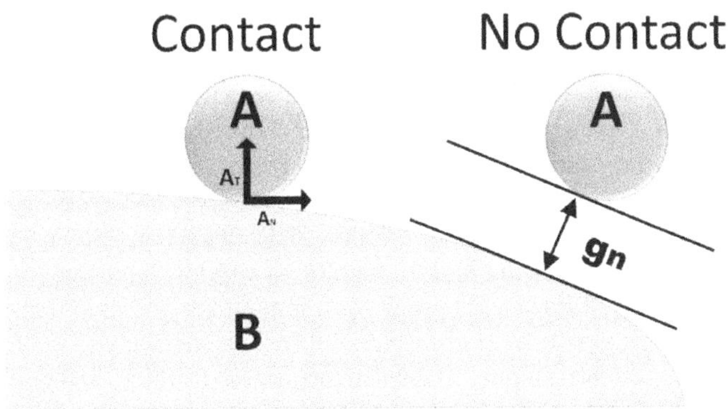

FIGURE 3.2 Schematic diagram of sliding contact between two bodies [10].

with industrial applications documented in the literature. The chapter closes with a concise articulation of the advances in smart surface technologies pertaining to challenges and the directions for future work in the area [11].

3.2 FUNDAMENTALS OF TRIBOLOGICAL BEHAVIOR AND SURFACE ENGINEERING

An understanding of the interaction of moving surfaces, coatings, and friction and wear control in the components of machines is based on the principles of tribology. The performance of almost every machine component depends on the component's ability to retain the surfaces of mechanical contact and accompanying reactions without deterioration. The tribological parameters of friction, wear, and lubrication constitute the principal attributes of non-degradation performance. The condition of the surfaces and interfacial wear films defines the topography, adhesion, and hardness, all of which influence the films under which the tribological parameters operate. In the engineering system, the contact surfaces deform under load, concentrating the mechanical stress which, in turn, leads to localized heating, friction, and wear. Load redistributing and minimizing tribological contact of sharp asperities to the surfaces is the goal of modern tribology. The design and production of hybrid machine components rely on the integration of surface engineering with systems and mechanical design to ensure that the bulk of the material retains its original properties. Redistribution of contact loads minimizes sharp apex asperities, conserving the operational load. Apart from material characteristics, such as hardness and elastic modulus, the mechanical performance of coatings is influenced by adhesion strength, arrangement of residual stresses, uniformity of microstructure, and other factors. It is possible for a coating to become overly hard, to remain weakly adhered, and consequently detach during cyclic loading. A weakly adherent coating that is overly soft will, however, undergo permanent deformation such that the substrate will eventually be revealed. Therefore, the balanced design of tribologically effective coatings in toughness, hardness, and adhesion is fundamental

[12]. Figure 3.3 is a summary of the many simulation methods that are commonly used in tribology to study how surfaces interact, wear, and lubricate each other.

Coated surfaces in frictional interaction undergo the dynamic processes of adhesion, abrasion, plowing, and deformation. When sliding occurs, microcontact heating at asperities may produce softening at the contact interface or phase transitions. For instance, high-temperature coatings form oxidized layers that act as self-lubricating barriers, reducing friction and enhancing stability. Conversely, the hard and soft phase distributions in metallic or ceramic nanocomposites control the transfer films that influence the wear. Titanium nitride and diamond-like carbon coatings mitigate adhesive wear characterized by low surface energy and elevated hardness, and solid lubricants such as sulfur, tungsten disulfide, and graphite layers markedly reduce friction to lower than 0.05 in optimal conditions [13]. Figure 3.4 is a diagram showing how molecular dynamics simulation works, atomic-scale interactions, frictional behavior, and wear mechanisms at a tribological contact interface.

In materials design, surface texture and microstructure are as important as materials chemistry. Incorporating nanostructures or gradient architectures improves load-bearing capacity as well as offers energy dissipation pathways that postpone crack initiation. Refining grains increases hardness via the Hall–Petch mechanism, as smaller grains are able to restrict dislocation movement [14]. On the other hand, amorphous coatings such as DLC have high elastic recovery and reduced adhesion because of the lack of grain boundaries. Interface engineering can be found in multilayered coatings, where alternating hard and soft layers reduce crack propagation by deflection or arrest microcracks before reaching the substrate. In tribological systems, the external environment has a great impact on coating performance.

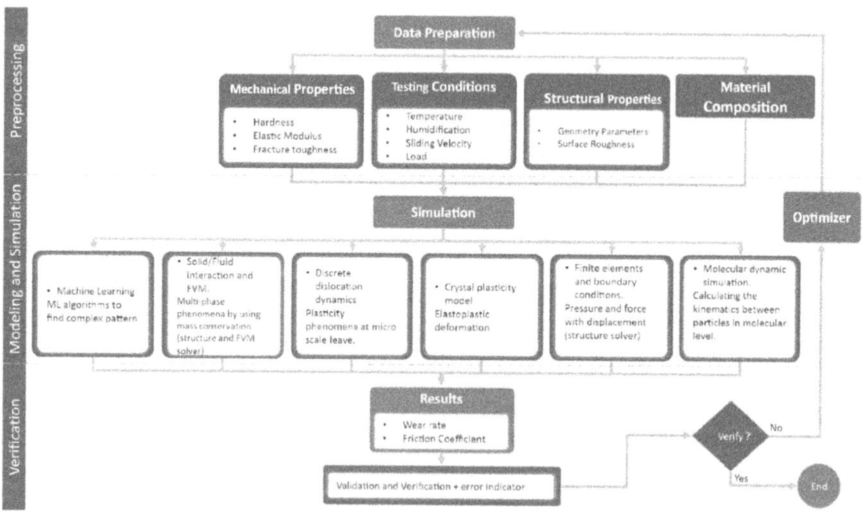

FIGURE 3.3 Overview of different simulation techniques commonly used in tribological-related applications [10].

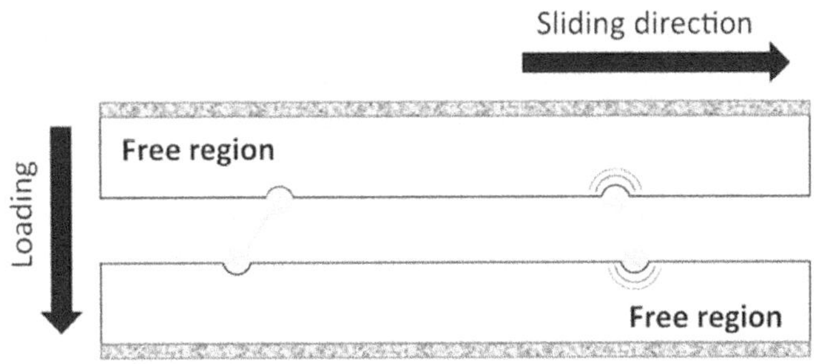

FIGURE 3.4 Schematic representation of molecular dynamic simulation of a tribo-contact [10].

Temperature, humidity, oxygen content, and corrosive species all contribute to surface changes that can improve or degrade coating performance. For example, high-temperature coatings need to be chemically stable to avoid the spallation of oxides, while in humid environments, hydrophobic coatings reduce capillary adhesion and thus friction. The integration of tribological and corrosive processes, or tribocorrosion, poses considerable challenges in marine, energy, biomedical, and other fields. In these situations, the passive films created under the conditions of sliding may experience simultaneous mechanical removal and electrochemical dissolution, thus requiring coatings with both chemical inertness and mechanical toughness [15, 16].

Assessment of the tribological behavior of these systems combines both experimental and analytical methods. The quantification of friction and wear and the assessment of these conditions in the pin-on-disc, ball-on-flat, and reciprocating wear tests are performed in controlled settings. In these tests, contact stresses and sliding speeds are set to values approximating the service conditions. The identification of the principal wear mechanisms (abrasion, adhesion, and fatigue) in a given sample is performed using surface characterization techniques such as scanning electron microscopy (SEM), atomic force microscopy (AFM), and energy-dispersive X-ray spectroscopy (EDS). Localized nanoindentation tests provide data for the construction of profiles of hardness and elastic modulus, and these profiles are requisite for the analysis of mechanical gradients present in multi-layer coatings. In tribo-chemical reactions, detection of graphitization in diamond-like carbon coatings or oxidation of metallic coatings is performed using Raman spectroscopy and X-ray photoelectron spectroscopy (XPS) [17]. Real-time mechanisms are best captured when tribological testing is performed in situ under controlled atmospheres or elevated temperatures. Crack onset and delamination during sliding are captured using acoustic emission and high-speed cameras. Moreover, integrated electrical and thermal sensors on test rigs are able to provide quantitative data on the evolution of contact temperatures and the rate of transfer between materials. This level of detail enables the concentration of microstructural design and performance, thereby establishing robust relationships between process, structure, and properties [18].

The exploration of new computational models is changing the understanding and prediction of tribological phenomena. Interfacial shear behavior of systems involving solid lubricants and thin oxide films can be explored at the atomic level using molecular dynamics simulations. Finite element modeling (FEM) is capable of microscale analysis by simulating stress distribution and contact deformation of multilayered coatings. Combined tribo-mechanical and thermal simulations can prognosticate failure modes under cyclic loading and inconsistent temperature variations. In addition, the use of machine learning can uncover new hidden relationships in large experimental datasets. For instance, machine learning predictive models can determine the friction and wear of coatings with varying compositions, porosity, and deposition rates. Materials discovery will be even more enhanced with the use of these data-driven methods. The most basic problem in tribology research is the reproduction of the multifaceted, real-world scenarios. Most laboratory tests tend to use simplified configurations with limited vibration, contamination, and transient loading. Hence the need for the multiscale study that combines experimental studies, theoretical models, and machine learning for data-driven optimization. This will produce the most reliable method for predicting performance [19].

Coatings calibrated for Industry 4.0 not only need to perform mechanically and chemically, but they also need to show stable performance across a range of operational conditions, including those governed by sensor feedback systems. There is a need to focus on advanced surface treatment to meet these needs through thoughtful and intelligent design. The latest coatings are not just passive barrier coatings. They actively participate in energy dissipation, self-repair, and communication within a mechanical system. By mastering the tribological fundamentals of a system—down to the asperity contacts and wear phenomena on macro scales—innovative researchers can optimize the performance of a system by adjusting the coatings' chemical and structural attributes. The next level of the coatings' performance will come from the merger of the science with digital tools of manufacturing, whereby coatings are designed, deposited, and even optimally adjusted on the fly by complete autonomy in the systems. Driven by a combination of tribology, materials informatics, and digitally controlled manufacturing, self-adaptive surface technologies are possible. Coatings are expected to shift from protective multifunctional coatings that meet specific needs to intelligent coatings that monitor and adjust to wear in real time and provide multifunctional surface systems [20].

3.3 EVOLUTION OF SURFACE TREATMENT TECHNOLOGIES

The development of surface treatment technologies demonstrates the ongoing attempt to enhance the performance, effectiveness, and longevity of materials subjected to aggressive operational conditions. Initially, surface treatments were purely chemical or mechanical, which altered the near-surface composition, offering only the most basic protection against corrosion and wear. The most primitive industrial applications of these treatments involved the deposition of metals or oxides through electroplating and anodizing to improve aesthetics and protect against corrosion. While suitable for some applications, these methods suffered from weak adhesion, environmental toxicity of the plating electrolytes, and loss of control over the microstructure. Proliferating industrial needs fostered new methods that, besides surface chemistry alteration, could control micro- and nanoscale mechanical, thermal, and tribological properties of the surface [21].

The development of vacuum techniques like physical vapor deposition (PVD) and chemical vapor deposition (CVD) marked a significant shift in the development of advanced surface technologies. Within the PVD family of techniques which include magnetron sputtering, cathodic arc deposition, and pulsed laser deposition, the ability to design hard, adherent coatings with microstructures of fine grain size and precise stoichiometry became possible. Coatings of titanium nitride (TiN), titanium aluminum nitride (TiAlN), and chrome nitride (CrN) became industrial benchmarks for cutting tools and aircraft components for their hardness and oxidation resistance in comparison to electroplated coatings, and which withstood friction at higher temperatures. CVD advanced the ability to design chemically controlled conforming coatings for surfaces of varying complexities, relying exclusively on gaseous precursors to produce dense films of engineering ceramics like silicon carbide and diamond-like carbon. These innovations in surface engineering shifted the focus from surface coatings of basic protection to custom-designed thermally and chemically engineered surfaces [22]. Figure 3.5 is a diagram that shows how smart coatings work by responding to outside factors like heat, stress, light, and pH, which lets the surface adapt and regulate itself.

Building upon the advancements in the understanding of surface phenomena, the development of hybrid and thermally oriented methods of deposition arose due to the limitations placed by PVD and CVD when coating large and irregularly shaped components. The thermal spray methods, which consist of plasma spray, high-velocity

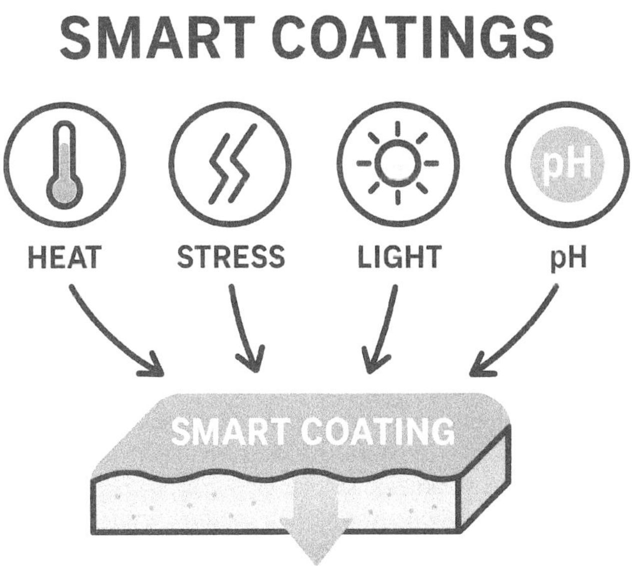

FIGURE 3.5 Schematic illustration of stimuli-responsive smart coating systems activated by heat, stress, light, and pH variations.

oxygen fuel (HVOF), and cold spray, are able to perform the deposition of thick coatings comprising metals, ceramics, or composites, as the coating is formed and solidified upon the rapid propulsion of molten or semi-molten particles toward the surface, resulting in dense and mechanically strong solidified coatings produced by interlocking. The thermal spray coating was able to provide to the aerospace industry the turbine blades, and to the industry of hydraulics the cylinders and wear-resistant rolls in which the combination of toughness and hardness is of the utmost importance. The ability to control the microstructure was even more advanced in suspension plasma spraying and solution precursor plasma spraying, resulting in performance enhancements due to grain-boundary strengthening and crack deflection [23].

The sol–gel method has also become a prominent method for making low-cost oxide and hybrid coatings. Additionally, making oxide and hybrid coatings with this method can be carried out at low processing temperatures. This process goes through the hydrolysis and condensation of metal alkoxides and/or inorganic salts. This process ends with a gel state that can then be spin-coated, dip-coated, or sprayed onto a substrate. This gel then turns into either amorphous or crystalline oxide films after curing. The films have excellent optical and chemical properties. One of the types of hybrid sol–gel coatings that have become very essential is those with organic polymers and/or nanoparticles for multifunctional applications and corrosion protection. These also have a low thermal budget and chemical tunability, which is important for use with light alloys, plastics, and other temperature-sensitive materials. Sol–gel coatings with inhibitors, pigments, or self-healing mechanisms have expanded their use to smart coatings that respond to sol–gel environmental triggers. The smart coatings respond to environmental triggers with the use of ultraviolet light layers [24]. Integrating laser and plasma techniques takes localized treatment without bulk distortion to a higher level. For a given surface treatment that involves laser melting and alloying, a high-energy laser beam is focused to melt a thin layer of a laser-compatible substrate, or preplaced powder is melted, forming a metallo-bonded coating. Depending on the treatment, those coatings may have fine-grained or even amorphous microstructures. For thermal (heat) treatments of nitriding and carburizing, the plasma as a nitrider and carburizer developed surface layers of hard nitrides and ceramic carbides to improve the surface composition of steel and titanium alloys. These layers provide excellent resistance to fatigue failure and abrasive wear. More recent improvements in plasma-enhanced chemical vapor deposition (PECVD) systems have allowed the low-temperature deposition of hard films on heat-sensitive substrates, like polymers or aluminum alloys, extending the application of PECVD even further [25].

In recent decades, atomic layer deposition (ALD) has emerged as one of the most sophisticated coating techniques. ALD excels based on the ability of self-limiting surface reactions to provide several layers of precisely controlled coating thickness. ALD creates smooth, conformal coatings even on complex three-dimensional (3D) architectures. This unrivaled conformality has made ALD crucial in microelectronics, biomedical implants, and energy systems where coatings have to be uniform. The ability to deposit diffusion barriers and protective layers made of several-nanometers-thick alumina, titania, and hafnia has changed the corrosion and wear protective strategies at the nanoscale. ALD, in combination with assisted plasma and other hybrid techniques, offers the unique ability to fabricate nanolaminates and engineered gradient coatings with customizable mechanical and chemical properties. The evolution of surface treatment technologies has, in part, been the result of compliance with environmental regulations and the sustainable development agenda. The corrosion protective, chromate conversion coatings have been replaced with less toxic and less environmentally dangerous alternatives, such as trivalent chromium and phosphate. The advancements in cold spray technologies, which deposit particles below their melting points, have certainly improved energy efficiency during deposition, minimizing oxidation and residual stress in the deposits. With the ability to repair components and provide coatings for additive manufacturing, cold spray coating technologies support the principles of circular economy by extending the life of components through material reuse [26].

The integration of materials science and digital engineering has led to the development of intelligent surface processing. Today's deposition systems rely on real-time monitoring systems and closed-loop controls to automatically adjust temperature, gas flow, and power based on feedback through optical emission spectroscopy or pyrometry. This integration leads to continual quality, reproducibility, and process optimization. Moreover, the use of data-driven machine learning techniques to map process-property relationships, predict coating defects, and microstructure optimization adds to the change from empirical process control to predictive adaptive manufacturing. The development of surface engineering from simple systems of surface finishing to advanced systems of smart digitally controlled surface coating systems provides ample evidence of progress. Each surface engineering technology has advanced the field: electrochemical plating, laser-assisted deposition, and ALD added new capabilities to the construction of functional surface materials. Next-generation smart coatings rely on these capabilities. The coatings are discussed in the following sections and include nanostructuring, digital self-healing, and adaptability to change the performance of surfaces defined within the Industry 4.0 framework [27].

3.4 NANOCOATINGS AND NANOSTRUCTURED SURFACE MODIFICATIONS

Advancements in nanocoating technologies have shifted surface engineering in new directions due to improvements in material performances that begin at the nanoscale. The significant attributes of nanocoatings include their unique fine-structured grain arrangement, their considerable surface-to-volume ratio, and the additional functionalities of superhardness, self-cleaning, catalytic properties, and more. The use of nanostructures, either in the form of discrete nanoparticles, layered nanocomposites, or continuous nanophases, offers unprecedented engineering of the mechanical, thermal, and electrochemical surface properties. TiN and DLC, along with molybdenum disulfide (MoS_2) and alumina–zirconia nanocomposites, serve as examples of "super" engineered systems that have improved tribological and oxidation resistance through grain boundary strengthening and quantum confinement. The reduction in the size of the micro-coating structure not only improves hardness but also enhances adhesion and toughness. Fine-grain interfaces are created, which significantly stop crack growth [28].

Different methods, such as atomic layer deposition, magnetron sputtering, pulsed laser deposition, and sol–gel routes are used to synthesize nanostructured coatings. These methods provide the means to finely adjust the layer thickness and composition to the atomic level, and customize the surface functionalities. In the biomedical field, nanostructured hydroxyapatite and zirconia coatings have osseointegrated and decreased bacterial adhesion, while energy systems utilize nano-TiO_2 and ZnO coatings for enhanced photocatalytic and protective abilities. The addition of 2D materials like graphene and MXenes in corrosion resistance and electrical conductivity research is groundbreaking, especially since graphene is ion impermeable and has extraordinary carrier mobility. Nanocomposite coatings with metal oxides and carbon nanostructures have excellent corrosion resistance due to the dense, self-passivating, and defect-tolerant barriers they form [29]. Figure 3.6 is a diagram of a multifunctional nanocomposite coating that shows how nanoparticles work together to improve mechanical strength, corrosion resistance, and tribological performance.

At the nanoscale, the incorporation of stimuli-responsive behavior at the surface is possible, where nano-sized inclusions respond to thermal, optical, or electrochemical stimuli, paving the way for the advanced intelligent smart coatings of the future. In addition, current advancements in molecular dynamics simulations in combination

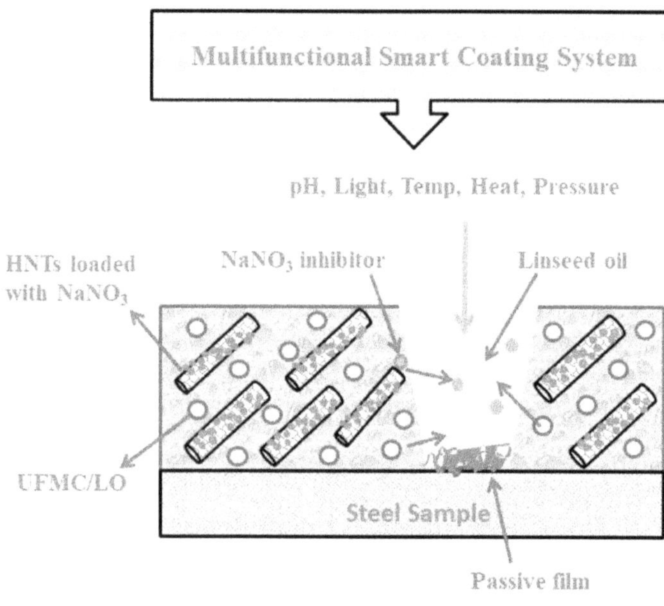

FIGURE 3.6 Schematic representation of a multifunctional nanocomposite coating [30].

with machine learning-based design open pathways for the prediction of interfacial energetics and growth dynamics at the atomic scale, furthering the rapid advancement of nanostructured film technologies. The need for high-performance surfaces in areas such as aerospace, microelectronics, and biomedical implants is likely to ensure that the functional design of nanocoatings continues to integrate multifunctional materials, the first of which will be bridging the structural strength with the functional versatility of the surface [31].

3.5 SELF-HEALING AND STIMULI-RESPONSIVE SMART COATINGS

Self-healing and stimuli-responsive coatings mimic adaptive biological and environmentally responsive systems. Self-healing materials are designed to regain functionality regardless of damage mechanically inflicted or caused by external temperature, pH, light, or electrochemical potential. Self-healing may be intrinsic, with the material matrix undergoing reversible rearrangement, chemical or physical, or extrinsic, with microcapsules, vascular networks, or nanoparticles with healing agents. Cracks or defects may form, and healing agents will be activated to polymerize, chemically react, or otherwise join to the damaged region to restore contour or solid form and resistance to corrosion [32].

The life of coatings under severe environmental conditions has been significantly enhanced by the microencapsulation of sol–gel or polymer matrix corrosion inhibitors, silanes, and epoxy resins. For example, cerium or zinc corrosion inhibitors, within nanocontainers, encapsulated, will migrate toward defects and repassivate and cease corrosion to exposed metal surfaces providing autonomous corrosion protection. Stimuli-responsive coatings use dynamic chemical bonds or phase-change materials and respond to and activate/disengage in the presence of external stimuli. Thermochromic materials form and are used as coatings where the temperature induces a color change or a shift in the balance of visible light transmittance through an electrochromic film. pH-sensitive and corrosive ion layers control corrosion activity and ion diffusion within the coatings [33]. Figure 3.7 is a diagram of a self-healing coating that shows how it may repair cracks on its own by releasing healing chemicals, which restores its protective function and surface continuity.

Hybrid coatings that self-heal and respond to stimuli are capturing interest in the aerospace and energy industries due to their long-term reliability and adaptability to varying environmental conditions. Employing nanoclay and graphene-based carriers further enhances responsiveness and diffusion control of active agents. Such coatings are positioned to dramatically decrease maintenance costs and increase safety, solidifying self-healing coatings as foundational elements for the next generation of sustainable and high-performance infrastructures and systems [34].

3.6 ADAPTIVE AND INTELLIGENT COATINGS

Adaptive and responsive coatings are the next step in the evolution of smart materials; all the workings of an integrated system of sensing, processing, and

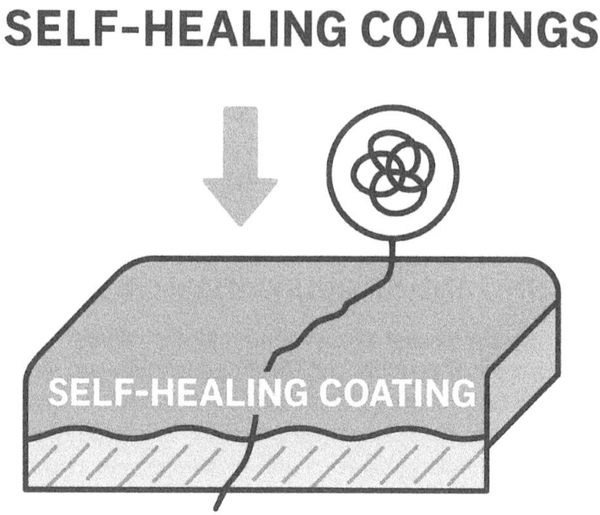

FIGURE 3.7 A diagram that shows how a self-healing coating works, with encapsulated healing agents or dynamic polymer networks automatically fixing microcracks and restoring the surface's integrity and protection.

actuating behavior are contained within the surface coatings of materials. While conventional coatings just passively resist degradation and protect the materials, smart coatings are capable of recognizing when the environment is changing and fully autonomously responding to keep the material within a predetermined performance envelope. Intelligent materials achieve this with embedded sensors, shape-memory frameworks, and functional nanostructures designed to actively interrelate with the environment. For instance, coatings with embedded piezoelectric or magnetostrictive phases can trigger and respond to adaptive phase morphing when the coating is stressed, strained, or placed within a magnetic field. In a similar fashion, adaptive coatings within tribological systems are used to modify the friction coefficients and lubrication characteristics of solid lubricants such as MoS_2, graphite, or adaptive oxides built to endure phase transformation. The turbine engine thermal barrier coatings turn adaptive as the engine undergoes thermal cycles. Yttria-stabilized zirconia (YSZ) and rare-earth zirconates in the engine undergo a microstructural, thermal phase transforming cycle. Photothermally responsive coatings are designed to thermally self-regulate in aerospace and solar applications. They turn adaptive as the engine undergoes thermal cycles. These coatings can respond to a predetermined shift in temperature, and in turn, photothermally responsive coatings can thermally self-regulate in aerospace and solar applications [35].

Smart coatings add more versatility with data integration along with feedback. Predictive maintenance is made possible with coatings containing microsensors and/or wireless systems that relay information regarding temperature, strain, and possible corrosion at a given monitoring point. This is a layer that is part of the so-called digital materials thesis where the surface can actively provide some intelligence and information. Real-time diagnosis and adaptive process optimization become possible with the integration of machine learning algorithms and materials of this type. Besides continuous protection, these coatings provide valuable information about the state of a component that is critical to minimizing downtime. This is a notable feature in high performance applications like biomedical implants, reactors, and turbines [36].

3.7 TRIBOCORROSION AND MULTIFUNCTIONALITY

In engineering, tribocorrosion describes the simultaneous actions of mechanical wear and electrochemical degradation of engineering materials in a corrosive mechanical working environment. Since mechanical wear-resistant and corroding-protective coatings tend to fail under different loadings, there is a need for multifunctional coatings. Designing nanocomposites and functionally graded coatings in modern surface engineering attempts to harmonize the tribological and electrochemical attributes of coatings. Systems such as CrN–TiN multilayers, WC–Co–Cr cermets, and DLC–metal oxide hybrids exhibit high hardness, low friction, and excellent corrosion resistance due to optimized phase interfaces and residual stress control [37]. Tribocorrosion-resistant nanocomposite coatings utilize disparate design to construct nanoscale microstructures that frustrate crack inception and electrochemical assault in a coating. Amorphous carbon or ceramic phases provide mechanical bulk, while self-passivating metallic matrices assist in electrochemical control. Functionally graded coatings enhance durability and offset the risk of delamination by distributing interfacial and compositional gradients in hardness, stress, or delamination. Fretting corrosion mitigation is achieved by TiN and hydroxyapatite (HA)-based coatings in biomedical implants, while Ni-based cermet coatings are used in the marine and energy sectors to withstand cavitation along with electrochemical corrosion [38].

Hybridization also allows for multifunctionality by adding solid lubricants, corrosion inhibitors, and hydrophobic elements in one matrix. This strategy permits coatings to cope with mechanical and chemical challenges while preserving structural integrity for long periods. New modeling approaches and in situ tribocorrosion tests demonstrate that adjusting certain microstructure–property relationships, primarily at the interface, prolongs the life of coatings. The integration of tribocorrosion-resistant coatings with predictive maintenance approaches follows Industry 4.0 standards by enabling monitoring of system performance and active failure avoidance [39].

3.8 INDUSTRY 4.0 AND DIGITAL INTEGRATION IN COATING TECHNOLOGIES

The beginning of Industry 4.0 has changed surface engineering due to the introduction of digitalization, automation, and optimized processes built on data. Coating technologies are beginning to advance from systems that are controlled empirically to systems built of intelligent sensors that are capable of real-time monitoring and adaptive control. Digital twins and cyber-physical systems of coating processes allow parameter–property interactions to be simulated before actual surface deposition. This ensures quality consistency across complex geometries and varied batch scales while minimizing trial and error. Cutting-edge sensors inside the coating chambers and measuring systems that analyze temperature, plasma intensity, and particle velocity feed data to algorithms that analyze machine learning, predicting coating attributes such as porosity, adhesion, and residual stress. The robotics integrated with PVD, CVD, and thermal spray systems enhance trajectory control and thickness control to reduce human error and material wastage. Also, the use of cloud-based data analytics for remote predictive maintenance of dispersed manufacturing units has become a standard practice [40]. The use of AI in coating design creates new opportunities in process innovation. Knowledge-driven optimization is fueled by digital integration. AI systems can define optimal process windows and material combinations to achieve specific advanced functionality. Such intelligent linkages result in shorter development cycles, improved sustainability, and greater surface engineering throughput, affirming its position as a digitalized pillar of advanced manufacturing ecosystems [41].

3.9 ADDITIVE MANUFACTURING AND COATING SYNERGY

Additive manufacturing (AM) has developed into the transformational technology that expands and complements the science of coating. The merging of AM and surface modification technologies enables engineers to construct devices with specially designed microstructures and configured surface functions and features. Manufacturing surface coatings after AM can ease rough surface textures and reduce the porosity and residual stress of surfaces. This advancement enables enhanced performance of the surface to withstand mechanical stress, fatigue, and corrosive elements. HVOF, plasma spraying, and laser cladding are some of the many techniques used to add protective layers that make hybrid manufacturing systems possible [42].

This fusion is of great importance in the aerospace and energy biomedical sectors. This is specifically the case when lightweight, high-strength, and corrosion-resistant surfaces become indispensable. The combo of laser cladding and AM also permits the design of functionally graded structures that possess a seamless graded coating. Additionally, cold spray can also repair or enhance additively manufactured parts without causing thermal distortion. Coatings can also decrease surface diffusion, high-temperature oxidation, and wear. Smart coating design integrated with AM opens endless possibilities for the creation of multi-material structures, enhanced and efficient material distributions, and optimized coating thickness through computational design and real-time monitoring. This approach signifies the convergence of design and material science with digital fabrication aiming to deliver configured, custom, and performance-enhanced components for advanced engineering systems [43].

3.10 ADVANCED CHARACTERIZATION AND MODELING APPROACHES

The ability to understand and optimize coating systems on various scales requires more refined and more advanced characterization and modeling approaches.

Focused ion beam–scanning electron microscopy (FIB-SEM), transmission electron microscopy (TEM), and atom probe tomography (APT) are some of the novel analytical techniques that allow for the nanoscale visualization of the interface chemistry and the evolution of defects. In situ tribological and electrochemical techniques provide real-time insights into how a coating behaves in real service environments. Synchrotron x-ray diffraction and neutron reflectometry are other tools used to measure residual stress and account for the crystallographic texture and other phase changes associated with various thermal or mechanical loading conditions [44].

Predictive simulation of coating stress migration, coating growth, and coating degradation kinetics mark the modeling aspects that the other techniques don't cover. Parameter optimization and simulation of erosion, wear, and heat transfer in coated systems are achieved through the use of finite element modeling (FEM) and computational fluid dynamics (CFD) along with other tools. Molecular dynamics and density functional theory (DFT) operate at the atomistic level to provide insights on bonding, diffusion, and other interfacial phenomena that impact adhesion and phase stability. Machine learning surrogate models integrate experimental and computational data to optimize coating designs and identify possible failure routes. Combining multiscale modeling and in situ diagnostics creates a digital coating design workflow that characterizes and forecasts performance under intricate service conditions. This hybrid experimental–computational strategy enhances the speed of innovation while ensuring dependability and reliability [45].

3.11 SUSTAINABILITY AND CIRCULAR ECONOMY IN COATING DESIGN

Resource efficiency, recyclability, and safety in use define sustainability in modern coating design. Eco-friendly coating alternatives are available to replace electroplating and chromate conversion. Eco-friendly alternatives replacing electroplating and chromate conversion coatings include trivalent chromium, phosphate free, and water-based sol–gel systems. These processes eliminate toxins and decrease newly generated waste while achieving high performance. Moreover, cold spray and laser techniques are designed to conform to low-energy, solvent-free deposition and green manufacturing standards. Within circular economy frameworks, we reduce, reuse, repair, and remanufacture coated components to extend their life cycles and reduce material demand. The coatings designed for easy removal or refurbishment allow performance-based value retention. Hybrid self-healing and corrosion-inhibiting coatings reduce maintenance frequency, extend cycles of harmful resources, and lower detrimental overall impacts. The use of bio-based coatings, inspired by renewable resources and created from non-petrochemical-derived polymers and opaque nanoparticles, are replacements for a coated material derived from harmful petrochemical polymers [46].

Another border of sustainability is energy-efficient processes. Compared to traditional thermal techniques, microwave and induction-assisted coating methods not only save on processing time but also save on energy. Life cycle assessment (LCA) tools are used more and more to capture the environmental impact of coatings and processes and help industries find greener alternatives. The combination of digital twin and AI-based optimization tools also helps sustainability by reducing the need for trial and error and by reducing waste of raw materials. Sustainable surface engineering is also (and more importantly) about the coatings harmonizing with the principles of resource circularity. Sustainable surface engineering can incorporate eco-design, the use of renewables, and smart trackers. Coatings will deliver on cutting performance while also being resource responsible. These will become the building blocks of resilient and eco-sustainable manufacturing systems [47].

Cross a range of different engineering fields, coating technologies can enable new surface and structural analysis capabilities. They provide surface roughness metrics, mechanical integrity property assessments (including tensile, compressive, and flexural strengths), and solar energy systems performance improvements via optical and thermal property modifications. Furthermore, coatings facilitate accurate simulations of fluid flow within complex channel networks and transitions (pipes and elbows). In the case of precision-engineered systems, such as ergonomically sensitive seating systems, the presence of functional coatings improves load mitigation and structural response predictability, resulting in greater durability, optimized design, and reduced error in multiphysics simulations [48–57].

3.12 CONCLUSION

Coating technology has come a long way; what started as covering surfaces in metal has advanced to printing surfaces that change as they are used. Coating surface technologies have redefined the basics of the baffling functions surfaces can perform. Through the evolution of digital technologies, coating surfaces have become integral to the pillars of the Fourth Industrial Revolution, from predictive data to 3D printing to the active use of self-healing surfaces and nanostructure design. Talent and the proper technology to utilize it have become the primary means to achieve advanced coating surface design. Employing advanced coating surfaces should always consider the diminishing returns of a global economy. The new frontier of coating surfaces will combine and multiplex with self-repair, and active surfaces will achieve a completely new level of coating design. Coatings will achieve a new, unprecedented level of active design. They will self-heal and actively communicate from a digital interface with a user, making it to the advanced level of surface design and production.

REFERENCES

1. Holmberg K, Erdemir A. Influence of tribology on global energy consumption, costs, and emissions. *Friction*. 2017;5(3):263–284.
2. Van der Zwaag S, et al. *Self-healing Materials—An Alternative Approach to 20 Centuries of Materials Science*. Springer; 2016.

3. Kumar S, et al. Smart coatings for intelligent manufacturing. *Surf Coat Technol.* 2022;439:128461.
4. Nosonovsky M, Bhushan B. Green tribology: principles, research areas, and challenges. *Philos Trans R Soc A*. 2010;368(1929):4677–4694.
5. Lee J, Bagheri B, Kao HA. A cyber-physical systems architecture for Industry 4.0-based manufacturing systems. *Manuf Lett.* 2015;3:18–23.
6. Zhang Y, et al. Machine learning for materials informatics in coating technologies. *Mater Today Commun.* 2023;34:106175.
7. Li J, Zhou P, Attarilar S, Shi H. Innovative surface modification procedures to achieve micro/nano-graded Ti-based biomedical alloys and implants. *Coatings.* 2021;11(6):647.
8. Vepřek S, Vepřek-Heijman MGJ. Industrial applications of superhard nanocomposite coatings. *Surf Coat Technol.* 2020;397:125965.
9. Grill A. Diamond-like carbon coatings: state of the art and future trends. *Surf Coat Technol.* 2021;429:127985.
10. Kurdi A, Alhazmi N, Alhazmi H, Tabbakh T. Practice of simulation and life cycle assessment in tribology—A review. *Materials.* 2020;13(16):3489.
11. Blaiszik BJ, et al. Self-healing polymers and composites. *Annu Rev Mater Res.* 2010;40:179–211.
12. Hao J, et al. Temperature-adaptive lubricants for advanced tribological systems. *Tribol Int.* 2023;183:108385.
13. Kurdi A, Alhazmi N, Alhazmi H, Tabbakh T. Practice of simulation and life cycle assessment in tribology—A review. *Materials.* 2020;13(16):3489.
14. Yin Y, et al. Nanostructured coatings for friction and wear reduction. *Prog Mater Sci.* 2022;130:101006.
15. Zhao H, et al. Graphene-reinforced composite coatings: a review. *Appl Surf Sci.* 2021;547:149170.
16. Bhushan B. Biomimetic tribology: nature inspired surfaces for green engineering. *Philos Trans R Soc A*. 2009;367(1893):1445–1486.
17. George SM. Atomic layer deposition: an overview. *Chem Rev.* 2010;110(1):111–131.
18. Varshney PK, et al. Advances in self-healing and adaptive coatings. *Prog Org Coat.* 2024;195:107011.
19. Liu Y, et al. Magnetorheological smart coatings for adaptive damping applications. *Smart Mater Struct.* 2022;31(8):085017.
20. Raheem A, et al. Digital twins for surface coating process optimization. *J Manuf Syst.* 2023;67:234–245.
21. Sundararajan S, et al. Machine learning-assisted prediction of coating tribological properties. *Tribol Lett.* 2024;72(1):9.
22. Ghosh P, et al. Reinforcement learning for autonomous coating process optimization. *J Intell Manuf.* 2025;36:1149–1163.
23. Padture NP, et al. Thermal barrier coatings for gas-turbine engine applications. *Science.* 2002;296:280–284.
24. Erdemir A, Donnet C. Tribology of diamond-like carbon films. *J Phys D Appl Phys.* 2006;39:R311–R327.
25. Huang S, et al. Self-cleaning coatings for renewable energy devices. *Prog Org Coat.* 2023;182:107646.
26. Deen KM, et al. Bioactive and antibacterial coatings for orthopedic implants. *Mater Sci Eng C.* 2022;138:112491.
27. Rao SS, et al. Sensor-enabled coatings for predictive maintenance in Industry 4.0. *IEEE Access.* 2024;12:57122–57136.
28. Blau PJ. *Friction Science and Technology: From Concepts to Applications.* 3rd ed. CRC Press; 2021.

29. Stachowiak GW, Batchelor AW. *Engineering Tribology*. 5th ed. Elsevier; 2022.
30. Habib S, Khan A, Nawaz M, Sliem MHR, Shakoor RA, Kahraman R, Abdullah AM, Zekri A. Self-healing performance of multifunctional polymeric smart coatings. *Polymers*. 2019;11(9):1519.
31. Meng Y, et al. Recent advances in tribology for surfaces and coatings. *Friction*. 2023;11(2):200–218.
32. Holmberg K, Matthews A. *Coating Tribology: Properties, Techniques, and Applications in Surface Engineering*. 3rd ed. Elsevier; 2020.
33. Kim SH, Asay DB, Dugger MT. Nanotribology and MEMS. *Nano Today*. 2007;2(5):22–29.
34. Shakhvorostov D, et al. Wear mechanisms of nanocomposite coatings under sliding conditions. *Wear*. 2023;526–527:204942.
35. Kato K. Classification of wear mechanisms/models. *Proc Inst Mech Eng J J Eng Tribol*. 2002;216(6):349–355.
36. Neville A, et al. Tribocorrosion understanding: from wear mechanisms to synergistic effects. *Tribol Int*. 2022;178:108032.
37. Tasan CC, et al. Microstructural design for improved damage tolerance in coatings. *Prog Mater Sci*. 2021;118:100772.
38. Vadiraj A, et al. Multi-scale modeling of tribological interfaces. *Comput Mater Sci*. 2023;224:112187.
39. Sharma A, et al. Machine learning-based prediction of friction and wear behavior in composite coatings. *Mater Des*. 2024;242:111065.
40. Verma PK, Vasudev H, Kumar V. Dental implant materials and designs and failure during implant placement and loading: A review. *Adv Manuf Process*. 2025:138–150.
41. Verma PK, Vasudev H, Kumar V. Bioactivity and corrosion performance of low-velocity oxy-fuel-sprayed hydroxyapatite-based coatings on a surgical-grade stainless steel. *Trans Indian Inst Met*. 2025;78(8):185.
42. Verma PK, Vasudev H, Kumar V. Effect of electro-corrosion behavior of different grades of stainless steel used in bio-implants: A review. *Adv Manuf Process*. 2025:269–283.
43. Verma PK, Vasudev H. Importance of hydroxyapatite coatings in improving new bone osteogenesis: A review. In: *Advanced Manufacturing Processes*. CRC Press; 2025. p.284–299.
44. Dutta V, Thakur L, Singh B, Vasudev H. A study of erosion–corrosion behaviour of friction stir-processed chromium-reinforced NiAl bronze composite. *Materials*. 2022;15(15):5401.
45. Singh M, Vasudev H, Kumar R. Corrosion and tribological behaviour of BN thin films deposited using magnetron sputtering. *Int J Surf Eng Interdiscip Mater Sci*. 2021;9(2):24–39.
46. Sharanabasava H, Raviprakash M, Prasad CD, Ramesh MR, Phanibhushana MV, Vasudev H, Kumar S. Microstructure, mechanical and wear properties of SiC and Mo reinforced NiCr microwave cladding. *Adv Mater Process Technol*. 2023;10:3620–3633.
47. Singh M, Vasudev H, Singh M. Surface protection of SS-316L with boron nitride based thin films using radio frequency magnetron sputtering technique. *J Electrochem Sci Eng*. 2022;12(5):851–863.
48. Prashar G, Vasudev H, Thakur L, Bansal A. Performance of thermally sprayed hydroxyapatite coatings for biomedical implants: A comprehensive review. *Surf Rev Lett*. 2023;30(1):2241001.
49. Vasudev H. Wear characteristics of Ni-WC powder deposited by using a microwave route on mild steel: Microwave cladding of Ni-WC. *Int J Surf Eng Interdiscip Mater Sci*. 2020;8(1):44–54.

50. Parkash J, Saggu HS, Vasudev H. A short review on the performance of high velocity oxy-fuel coatings in boiler steel applications. *Mater Today Proc.* 2022;50:1442–1446.
51. Singh P, Vasudev H, Bansal A. Effect of post-heat treatment on the microstructural, mechanical, and bioactivity behavior of the microwave-assisted alumina-reinforced hydroxyapatite cladding. *Proc Inst Mech Eng Part E J Process Mech Eng.* 2024;238(3):1060–1071.
52. Verma PK, Vasudev H, Singh G, Singh S. A review on the microstructural and biomedical properties of implants manufactured using additive manufacturing. *J Inst Eng India Ser D.* 2024;106:1–15.
53. Singh SP, Verma PK, Singh S, Kapoor M, Balu R. Microstructural characterization and electrochemical behaviour of HVOF sprayed WC-Co-Cr coatings on 316L boiler steel stainless steel. *Results Surf Interfaces.* 2024;16:100248.
54. Verma PK, Minhas AS, Singh P, Kumar S. Slurry erosion behaviour of thermal sprayed A12O3 and Cr2O3 coatings for turbine steels. *IAIP Conference Proceedings.* 2024;2986(1):020012.
55. Verma PK, Kumar V, Vasudev H. Bioactivity and corrosion analysis of thermally sprayed hydroxyapatite based coatings. *J Electrochem Sci Eng.* 2024;14(4):419–439.
56. Verma PK, Vasudev H, Kumar V. Biocompatibility and corrosion performance of HA-based LVOF sprayed and post-annealed coatings on SS 254 surgical-grade bio-implant material. *Ceram Int.* 2025;51(10):13207–13225.
57. Verma PK, Vasudev H, Singh G. Improving the corrosion performance of SS 304 stainless steel using HVOF-sprayed WC-Co-Cr coatings in chloride environments. *J Bio Tribo Corros.* 2025;11(4):1–13.

4 Optimizing Maintenance Strategies in Tribology Using Predictive Technologies and Real-World Implementations

Umesh Ghorpade, Shubhangi Suryawanshi, Pratik Rajendra Kachare, Sana Hassan, and Jeyaprakash Natarajan

4.1 INTRODUCTION

A new era known as Industry 4.0 has emerged as a result of the rapid advancement of technology, which has changed traditional industries. Connectivity, data-driven decision-making, automation, and customization in production systems are key components of the industrial revolution, which was initially implemented in Germany. Industry 4.0 is characterized by intelligent machines that self-configure, self-maintain, and self-organize, in contrast to previous models of industry. Data is the primary force behind this change since it allows both humans and machines to make prompt, well-informed decisions when vast amounts of data are available and combined with sophisticated analytics [1, 2].

Predictive maintenance, or PdM, has become the vital keystone of the manufacturing in this regard. To find patterns in behavior, spot irregularities, and anticipate failures before they happen, PdM uses statistical models, domain expertise, historical data, and machine learning techniques. PdM minimizes unscheduled downtime, lowers maintenance costs, and boosts productivity by foreseeing equipment failures.

It is this predictive methodology that has resulted in the development of larger ideas such as intelligent manufacturing or intelligent industry that focuses on the use of state-of-the-art technologies in order to optimize the performance of the industry [3–5].

Predictive strategies have become increasingly significant due to the increasing level of complexity of production systems. The operations of maintenance have to

 DOI: 10.1201/9781003741152-4

ensure reliability and effectiveness of a range of processes within the environment of extended manufacturing ecosystems and interrelated supply chains. In order to meet this need, real-time data collection and analysis are made possible by the Industrial Internet of Things (IIoT), which incorporates IoT technologies into industrial settings. IoT gives smart machines the ability to make data-driven decisions more accurately and consistently than humans when paired with big data and machine learning. The emergence of cyber-physical systems (CPS), the next generation of engineered systems that combine digital intelligence and physical processes. Data-to-information conversion, high-performance computing, cognition, and system configuration are among the layered architectures that underpin CPS, which was first presented by the U.S. National Science Foundation in 2006. These systems enable sophisticated modeling, monitoring, and control by offering abstract representations of machines and processes [6, 7].

Industry 4.0 intelligent maintenance teams are built on the convergence of PdM, IoT, and CPS. Despite the fact that data analytics, machine learning, and industrial digitalization have been the subject of numerous service systematic surveys, systematic reviews that concentrate exclusively on predictive maintenance are still required. This chapter fills this gap by proposing a structured taxonomy of PdM, which looks at its application domain, maintenance types, and guiding principles. It also examines monitoring and prediction models, identifies present issues, and suggests future lines of inquiry [8–10].

Because it integrates a variety of domains, including data acquisition, infrastructure, storage, distribution, security, and intelligence, PdM has grown in significance across multidisciplinary research fields. This section lays the groundwork for the discussion in this study and introduces the basic ideas required to comprehend PdM.

Between 15% and 16% of total manufacturing costs are attributed to maintenance activities, which constitute a substantial portion of industrial operating costs. Maintenance costs have a significant impact on many businesses that are not properly measured. This gap explains why it is important to study the latest technologies that can help save money and maximize output. In Industry 4.0, PdM can streamline maintenance processes and boost productivity in general [11, 12].

Industry 4.0 settings support the integration of sensors and other connected objects that allow the collection of large real-time volumes of data. Such data streams offer additional possibilities for forecasting the remaining useful life (RUL) of industrial assets. With this data, PdM develops maintenance plans not by time but according to the performance and conditions of the equipment over a period of time. Access to sufficient information at every step of the manufacturing process, say, is one of the key requirements for the successful implementation of PdM. When implemented correctly, PdM reduces costs, reduces downtime, and increases production and quality [13].

RUL is an issue that is prominent in engineering mechanics and automation. This notion lies central to the three major phases of prognostics and prognostics and health management (PHM): observation, analysis, and action. PdM is more or less associated with the stage of observation within PHM in which clever practices are observed in anticipating failures before they happen. Some scholars believe that

PdM is one of the key components of PHM, along with mean time to repair (MTTR) and equipment health monitoring (EHM) [14–16].

Four broad categories can be used to classify maintenance approaches (also see Figure 4.1):

- Corrective maintenance—Carried out following the discovery of a defect.
- Preventive maintenance—Planned at specified intervals.
- Predictive maintenance—Based on knowledge and data about conditions in real time, enabling failure prediction before it happens.
- Prescriptive maintenance—Aids in decision-making by offering suggestions for managing a team's particular failure events, which improves subsequent procedures [17].

PHM further divides maintenance methods into failure finding, corrective, fixed-interval preventive, and condition-based maintenance (CBM). Specifically, CBM employs mathematical models, algorithms, and sensor data to efficiently assess RUL [18, 19].

Additionally, PdM methodologies can be broadly categorized into three main approaches:

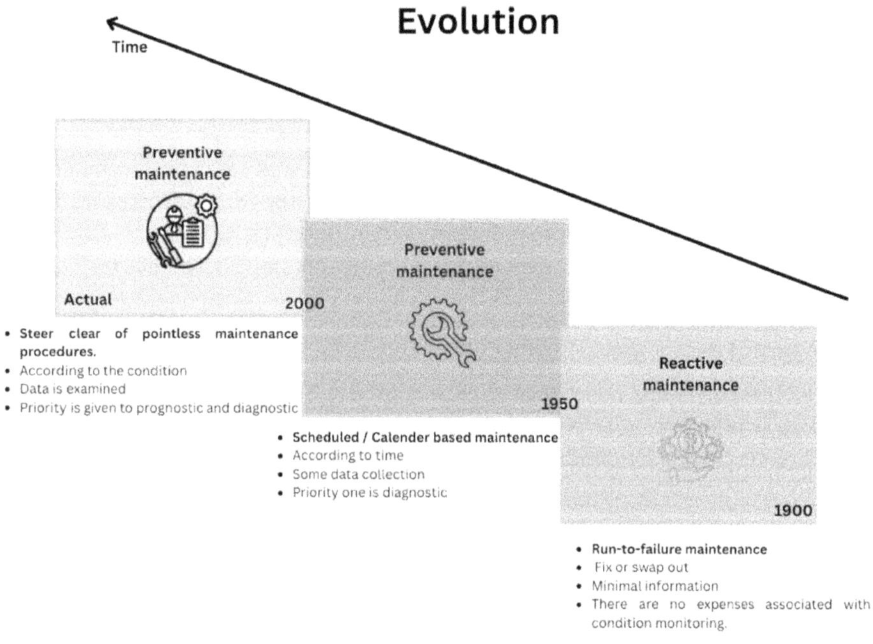

FIGURE 4.1 Evolution of maintenance activities and methods.

- Techniques based on physical models—Depend on mathematical modeling of component conditions, which necessitates accurate measurements and statistical methods for estimating the likelihood of failure.
- Knowledge-based approaches—Simplify physical models through the use of techniques like physiological or expert systems, which are frequently used as hybrid strategies.
- Data-driven approaches—The most popular in recent years, use machine learning algorithms, statistical analysis, pattern recognition, and artificial intelligence to forecast failures [20–22].

4.2 EMERGENCE OF PREDICTIVE MAINTENANCE IN TRIBOLOGY

Maintenance techniques have undergone significant change as a result of industrial revolutions. Industries have historically relied on reactive maintenance in which machinery was fixed only after it broke down. Even though this method is still frequently used in plants with limited resources, it often leads to expensive downtime and large losses. In light of the development of manufacturing technologies, industries slowly transitioned into preventive maintenance, i.e., the intervention performed with the purpose of extending equipment lifespan and reducing the possibility of unexpected failures. Nevertheless, since the intervention was forecast at predetermined times and not depending on the real state of the equipment, the preventive measures remained inaccurate [23].

Industry 4.0, or the Fourth Industrial Revolution, transformed the practice of maintenance and led to the notion of the connected and data-driven factory. PdM has emerged as an important initiative to ensure equipment reliability and efficiency. PdM is based on real-time data gathering, advanced analytics, and continuous machine data to forecast failure prior to its occurrence. This will reduce downtime and repair costs by increasing equipment life and reducing unnecessary interventions, improving sustainability [24].

PdM has some significant advantages in the specific area of tribology that deals with wear, friction, and lubrication of the surfaces in contact. It is now achievable to predict conventional tribological issues such as unnecessary wear, lubricant degradation, and bearing failures beforehand by integrating data analytics, sensor technologies, and intelligent diagnostic systems. This makes it possible to preemptively take preventive action, which will ensure the perfect lubrication schedules, early detection of atypical wear patterns, and greater machine reliability [25].

PdM has become possible in tribology due to a number of technology enablers across Industry 4.0 such as (also see Figure 4.2):

- Cyber-physical systems (CPS)—These systems also provide autonomous monitoring and diagnostic requirements through the combination of computational intelligence and physical machineries. CPS can monitor such variables as the temperature status, vibrations, and lubricant state in tribology and create a digital feedback loop between predictive models and real-time performance.

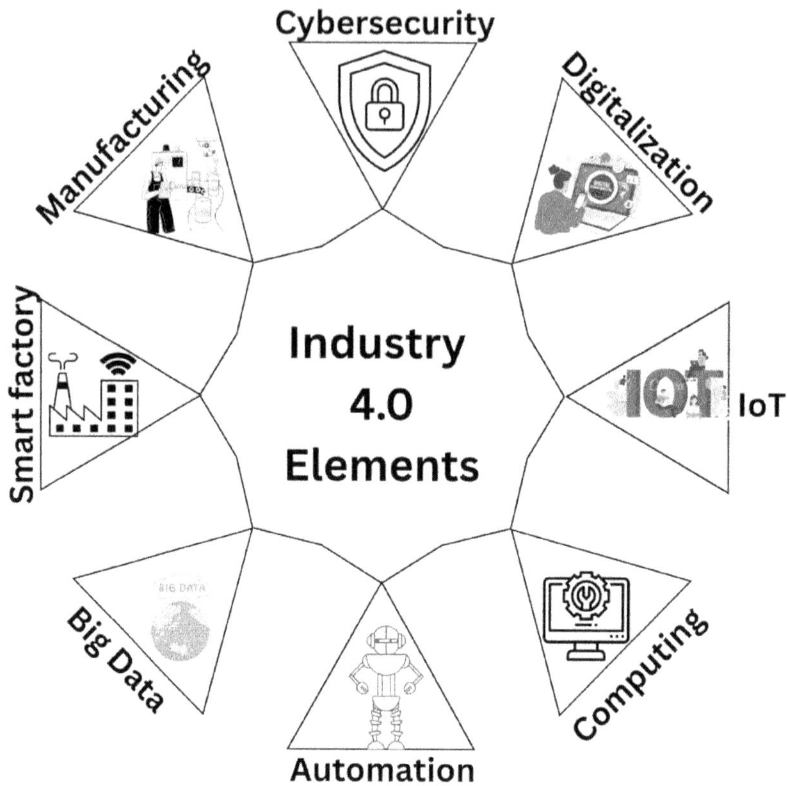

FIGURE 4.2 Elements of Industry 4.0.

- Industrial Internet of Things (IIoT)—IoT is used to connect machines, devices, and sensors in order to provide continuous data flow of tribological components such as bearings, gears, and lubrication systems. Such an instantaneous interconnection provides machine-to-machine communication and allows the extensive use of predictive wear and friction monitoring models [26].
- Big data analytics—By collecting and analyzing vast amounts of data streams of tribological systems, the wear patterns, frictional behavior, and lubricant performance are identified. This data is now processed by machine learning and deep learning methodology, which provide precise estimates of RUL of important parts.
- Digital twins—These are computerized models of real assets that take filtered data in constant feed. Digital twins allow simulating contact surfaces, wear rate, and lubricant behavior in diverse operating environments in tribology. This will enable improved design of the lubrication strategy, prediction of failures, and optimization of performances [27].

- Augmented reality (AR)—AR technologies offer interactive and immersive visualization devices to maintenance engineers. AR may help the technician lubricate, inspect correctly, and diagnose failures due to wear in tribological work.

4.2.1 Predictive Maintenance Life Cycle

An effective PdM strategy becomes successful when the structured life cycle is followed. This five-phase model makes sure that issues are addressed in a systematic manner through data science and machine learning.

4.2.1.1 Step 1: Understanding Project Requirements

The objective of the first stage is to understand the whole operation of the systems, business goals, and the existing challenges of the systems. It requires the close examination of the corresponding machinery or equipment, the working conditions, and possible implementation constraints. The primary physical variables to be measured at this stage are chosen, as well as the correct sensors and methods of installation. A list of possible failure modes is drawn in order to steer the later steps.

4.2.1.2 Step 2: Data Acquisition, Comprehension, and Preparation

- Data acquisition—The device has sensors that provide raw working data, stored in a database.
- Data comprehension—The data gathered is evaluated on the basis of quality, completeness, and relevance to the problem. This will only allow relevant data to be analyzed.
- Data preparation—This subphase is significant to success although it involves the heaviest work (approximately 70%–90% out of the whole effort). It includes data cleaning (processing errors, outliers, and missing values), combining several datasets, converting them into distinct useful forms, and engineering features. Proper data preparation makes the final model representative and reliable [28].

4.2.1.3 Step 3: Data Modeling

The foundation of predictive analytics is this phase; here, predictive models are created using the preprocessed data. Appropriate algorithms are chosen and adjusted based on the type of problem, such as clustering, regression, or classification. To determine which model provides the best predictive performance, several models may be tested and parameterized.

4.2.1.4 Step 4: Evaluation and Deployment

- Model evaluation—The accuracy, generalizability, and applicability of the developed models are evaluated. To prevent overfitting and maintain the model's applicability in real-world situations, the goal is to achieve the ideal balance between performance and robustness.

- Model deployment—The model is deployed in the proper format (software platform, embedded system, or cloud-based solution) after it has been validated. It is important to remember that deployment results will suffer from any flaw in the earlier phases (such as poor data quality or insufficient cleaning).

4.2.1.5 Step 5: Decision-Making

During this stage, model predictions are converted into workable maintenance plans.

- Identification—Clearly defining the decision problem including the various repair or prevention options and the corresponding expenses and downtimes in the first step.
- Action—The solution that saves the most money and time is chosen. Inventory spare parts and workforce availability are taken into account when planning maintenance schedules [29].
- Review—Continuous review is crucial because predictive maintenance is an iterative process. The system is improved through feedback from decisions that have been put into practice, which results in better planning and more effective interventions in subsequent cycles.

4.3 SENSOR TECHNOLOGIES FOR REAL-TIME TRIBOLOGICAL MONITORING

Analytical models of wear and friction or trial and error testing have historically been used to predict tribological failures in machine elements. However, these approaches are limited to particular operating environments and often fail to provide reliable insights. In order to overcome this limitation and enable intelligent decision-making and real-time condition monitoring, sensor technologies have revolutionized the field.

4.3.1 Monitoring Conditions with Sensorized Bearings

These days, sensors are commonly integrated into machine components to continuously monitor their condition. For example, encoder bearings with significant rotational counts and speed have been around for a while. Self-power sensing components and wireless communication are used in modern sensorized bearings to monitor parameters such as [30]:

- Rotational speed
- Operating temperature
- Vibration levels
- Bearing load

4.3.2 Tribotronic Devices and Feedback Control

Tribotronic control systems are found in many existing systems, including active clutches, electromagnetic bearings, and antilock braking systems (ABS). These include closed-loop feedback systems, sensors, and actuators. Such systems have the advantage of dynamically adjusting parameters based on sensor feedback rather than requiring extensive theoretical modeling. These concepts are currently being extended in controlled thrust pad bearings, tilting pad journal bearings, and seals where stiffness, stability, and efficiency adjustment is made possible by sensor input [31].

4.3.3 Technical Challenges in Sensor and Actuator Integration

One of the primary challenges in the development of active tribological systems is the creation of suitable actuation mechanisms. While mechanical adjustments can be made using electromechanical, hydraulic, or pneumatic systems, lubricants require chemical or energetic activation, such as electric or magnetic fields, to be stimulated. Furthermore, effective feedback control requires knowledge of the system's time constants. For example, due to thermal inertia and viscoelasticity, elastomeric seals and large bearing systems have a slow response. Accurate and timely data must be provided by the sensors in order to enable effective feedback mechanisms [31].

4.3.4 Sensors in the Context of Industry 4.0

The integration of sensors plays a major role in the Industry 4.0 vision, which emphasizes cyber-physical systems, connected devices, and smart factories. In this context, tribological sensors are very important for:

- Monitoring and diagnostics in real time
- Energy-efficient manufacturing solutions
- Predictive and electronic maintenance systems
- Integrating with digital platforms through the Internet of Things and wireless communication

The ability to share and transfer tribological data among devices enables digital control networks and physical machinery to work together without any interruptions, which in turn opens the pathways for automation and better resource utilization [32].

4.3.5 Lubricant Condition Monitoring

Nowadays, one of the main industrial objectives is to prolong the life of lubricants while reducing their harmful effects on the environment. A range of sensors is used to continuously monitor oil quality and lubricant degradation. Although the automotive and aerospace industries currently employ these systems, their industrial applications are still rather limited. The development of tiny, inexpensive, and extremely dependable sensors that can continuously and precisely monitor lubricant conditions is where the future lies [33].

4.3.6 Sensor Fusion and Machine Learning Integration

Tribological monitoring is about to be redefined by the combination of sensor fusion and machine learning. Machine learning algorithms can be used to analyze large datasets gathered from numerous sensors in order to find failure signatures and subtle patterns. It is feasible to differentiate between normal operation and pre-failure conditions using this technique. Combining supervised learning (for known conditions) and unsupervised learning (for unknown patterns) will make tribological process recognition more flexible and reliable [34].

The models' competency to estimate the RUL of parts like bearings throughout the period allows maintenance scheduling to be done proactively. Besides, machine learning systems fed by sensors will be necessary for predictive diagnostics in distant or safety-critical applications, like airplane engines and offshore wind farms, and in tribotronic components control through autonomy [35].

4.4 AI-DRIVEN PREDICTIVE MODELS

AI and ML have become revolutionary methods for predictive modeling in the field of tribology. AI-powered models are, in comparison to the conventional experimental methods that are frequently sluggish and subject to errors arising from faulty equipment or irregular test conditions, a quicker and more precise way of scrutinizing tribological properties. The use of machine learning algorithms in tribological research has grown significantly over the past 20 years; the growing dependence on AI for predictive modeling is evident from the fact that almost half of the published work in this field has been produced in the last five years alone.

Key tribological properties have been predicted by machine learning algorithms like the artificial neural network (ANN), support vector machine (SVM), random forest (RF), k-nearest neighbor (KNN), adaptive neuro-fuzzy inference system (ANFIS), convolutional neural network (CNN), and Bayesian model [36]. Their attributes include:

- Predicting the wear rate and friction coefficient under various operating conditions.
- Lubrication analysis in which machine learning models examine performance, corrosion rate, and viscosity under various lubrication regimes.
- Characterization of surface coatings, including estimation of frictional behavior, wear loss, porosity, hardness, and thickness.
- Condition monitoring and fault diagnosis, in which delimitation growth, bearing, and associated systematic life, defect identification, and remaining useful life are all predicted using AI techniques [37].

ANNs stand out among these techniques due to their high predictive accuracy (R-values near unity) in modeling complex nonlinear relationships. Similarly, decision trees (DTs) have demonstrated strong performance in fault identification classification tasks.

This review asserts computational intelligence as a basic necessity for the future of tribological research beyond the limits of traditional and purely statistical methods. The last two decades have been a rollercoaster of rising promise for AI modeling and optimization techniques in various fields. Despite the great success of ANNs in the treatment of highly complex and nonlinear tasks, an array of other algorithms like SVM, DTs, genetic algorithms (GAs), fuzzy logic, recurrent neural networks (RNNs), random forests (RFs), and ensemble learning (EL) have also been proven to be capable of providing promising solutions. The fusion of these resources provides the establishment of a versatile methodological framework that can be customized to different tribological applications [38].

AI methods can be applied to foresee forthcoming studies that would involve varying combinations of materials, test scenarios, and lubricant formulations along with the investigation of friction modifiers and nanoparticles' impacts on viscosity and wear. Nonetheless, one of the leading difficulties is the determination of the optimal algorithms since there is not one that can universally tackle all tribology issues.

One more major restriction is the lack of large, uniform datasets and the experimental test conditions that do not have proper control of uncertainty. In this case, the future work should be directed toward the making of open access data repositories and the appendices that cover experimental datasets, models, and documentation. Consequently, researchers will find it easier to benchmark and validate AI models, which will also contribute to increasing the reproducibility of their results [39].

In addition, the machine learning application in tribology is greatly dependent on the automation and optimization of data collection, which still comprises the major part of manual data collection. The growing use of nanoparticles in lubricants has not only increased the research area of tribology but also allowed researchers to explore the thermal-physical properties of nanolubricants and their influence on wear, friction, and lubrication performance as well.

Condition monitoring is a vital part of modern-day tribology as it allows for the constant affirmation of the proper operation of the machine by monitoring its speed, load, temperature, vibration, and electrical current. Moreover, state-of-the-art sensors are able to continuously follow the characteristics of the lubricant, and detect the presence of wear particles and degradation patterns, thus providing the earliest signs of component failure. Later, this data will be fed into AI algorithms that are able to predict the end of life of the machine, which results in preventive maintenance and reduced operating costs due to the less frequent occurrence of downtime. The aforementioned factors, namely, friction and wear, not only depend on the properties of the materials used but also on dynamic operating conditions such as surface roughness, coating, lubrication quality, and contamination, thus being the most significant tribological parameters that should always be monitored. These events can be modeled through machine learning techniques mainly due to their nonlinearity and temporal dependence [40].

ANN has proved to be a potent predictor in condition monitoring. Studies show that ANN models can capture the morphology and volume loss of components under wear with up to 85% more accuracy than other ML models. Alambeigi et al., for example, showed that an ANN trained with the Levenberg–Marquardt algorithm had a root mean square error (RMSE) of 4.21E–4 and R^2 value of 0.9912, confirming its high predictive ability in simulating the wear behavior of steel components [41].

In addition to laboratory validation, AI-driven condition monitoring has applications in industry. For example:

- Tool wear detection—Die and Zhu developed a real-time wear recognition system that allowed precise identification of drilling tool wear regions by utilizing additional drive signal data and image-based filtering techniques. Their approach reduced failures and increased lifespan [41].
- Transportation systems—Real-time limiting friction at the wheel–rail interface in trains has been predicted using AI models to increase safety and reduce wear-related failures.
- Wind power industry—AI and ANNs, in particular, aid in operation control, bearing deterioration prediction, and wind turbine design optimization. This can prevent unforeseen mishaps, reduce risks to those who work at significant heights, and decrease maintenance expenditures [42].

Physics-informed neural networks and hybrid AI models have also been able to prove their worth in the area of bearing failure tracking. The learning process was made more effective by the addition of physical laws, like monotonic temperature trends, which resulted in better predictions of component RUL. It guarantees that a prediction is in line with the already known degradation mechanisms. Hybrid techniques, like physics-informed temporal convolutional networks, marry machine learning with physical-based health indicators and deliver accuracy that is superior to conventional data-driven methods [43].

4.5 ROLE OF ADVANCED COATINGS IN MAINTENANCE OPTIMIZATION

Maintenance optimization is needed to improve the longevity and operational dependability of industrial components. Advanced surface coatings have been a major factor in predictive maintenance since they minimize wear, corrosion, and friction in very harsh service environments. This review discusses the technological advancements related to hard coatings, self-lubricating materials, smart sensor-embedded layers, and nanostructured films. Moreover, the integration of these technologies with AI and IoT platforms for condition-based maintenance plans, tribological data collection, and real-time diagnostics is also addressed. The review demonstrates the role of coatings in the Industry 4.0 ecosystem by presenting real-world applications in the manufacturing, aerospace, and biomedical sectors through a set of case studies.

The usual reactive maintenance methods cannot handle the situation caused by the rising industrial demand for equipment's uninterrupted operation, reliability, and cost-effectiveness, resulting in the occurrence of catastrophic failures. Continuous deterioration of the surfaces of highly efficient machines through the processes of wear, corrosion, and heat stress is still a significant concern. Advanced coatings developed at the micro- and nanoscales serve as the first line of defense against tribological deterioration. These coatings make possible predictive failure analysis and early anomaly detection, which are characteristic features of Industry 4.0 maintenance paradigms when coupled with sensors and data-driven models.

Wear on cutting tools and dies is considerably reduced due to the use of hard coatings such as TiN, CrN, and AlTiN, which offer superior hardness (more than 2000 HV) to cutting tool materials. They are perfect for machining and aerospace applications because of their low coefficient of friction and high resistance to oxidation. Low-shear layers from materials like MoS_2 and WS_2 enhance performance in dry or vacuum environments. Friction is further decreased by hybrid composites that combine ceramic matrices with graphite or PTFE. To improve durability, multilayer systems such as TiN/CrN or DLC/ZrN alternate layers of various modulus materials to prevent crack propagation and handle thermal stress. Ceramic and metallic overlays for turbines and industrial rotors are commonly deposited using high-velocity oxygen fuel (HVOF) coatings and atmospheric plasma spray (APS). In the oil and gas, power generation, and marine industries, coatings with anticorrosion, antifouling, or temperature-sensitive properties are essential.

Real-time information on stress, load, and wear progression is provided by coatings that have integrated strain gauges, piezoelectric components, and tribosensors. Orthopedic implants and smart turbines are using these more and more. Passive alerts and condition indications are made possible by materials that alter their characteristics in response to mechanical stimuli, pH, or temperature (such as color-changing polymers). IoT hubs receive sensor data streams, and AI models trained on past stress and wear data identify anomalies. Before symptoms appear, predictive algorithms notify users of upcoming failures.

The use of advanced surface coatings in maintenance strategies represents a radical shift in the methods of protecting, monitoring, and controlling industrial systems. Their application brings a significant increase in the reliability and lifespan of parts that work in extreme conditions; among the advanced coatings are those made of hard and self-lubricating materials, nanocomposites, and sensor-integrated smart films.

Smart coatings have recently made significant progress, and coatings have undergone a transformation in their role within predictive maintenance systems, going from being passive protectors to active participants. The addition of sensing capabilities, such as embedded piezoelectric elements or materials that react to temperature and pH, allows for continuous and real-time monitoring of the health of components. To avoid interruptions and maximize efficiency, the data generated by these smart coatings is being used with AI and machine learning models to predict malfunctions much earlier than they actually occur.

Advanced coatings have been applied to a range of products such as metal cutting tools, aerospace turbine blades, orthopedic implants, and heat exchangers, and the case studies reveal their broad industrial applicability and significant performance improvements. Furthermore, by reducing maintenance costs and frequency, these technologies contribute to the safety and sustainability of operations in various fields. The tribology, material science, IoT, and AI synergism will become a major driving force in the prediction of maintenance through the paradigms of predictive maintenance. Coatings will not only be viewed as protective films but also as pieces of intelligent and self-aware systems. The future of research is likely to be in the area of coatings with multiple functions like microdamage repair, environmental adaptation, and cooperation with cloud-based digital twins. Nevertheless, standardizing sensor data protocols, ensuring material compatibility, and reducing the costs of implementation remain challenges to be overcome. Continuous collaboration among government agencies, industry, and academia is essential to overcome these obstacles. It will be very important to improve deposition techniques, implement large-scale production of smart coatings, and train predictive models with real data.

In conclusion, advanced coatings are no longer just surface treatments but are rather becoming indispensable components of intelligent, adaptive, and data-driven maintenance ecosystems that optimize maintenance. Their increasing involvement is a big step toward the development of fully automated, highly reliable, and sustainable manufacturing processes.

4.6 FUTURE TRENDS AND RESEARCH OPPORTUNITIES

Predictive maintenance in industrial systems has a bright future ahead, but the realization of its potential depends on the successful tackling of a few significant challenges. The solution to these issues will create new paths for research and innovation, resulting in the development of smarter, self-reliant, and affordable maintenance methods. The upcoming paragraphs discuss possible future research directions and trends: Predictive maintenance is still suffering a lot from organizational and financial restrictions. Future studies need to concentrate on the creation of cost-effective solutions that are scalable and do not require imposing huge costs upfront. The use of modular sensors, pay-per-use services, and cloud-based platforms are some ways to solve the high-cost problem of predictive maintenance for large, medium, and small businesses. Moreover, the integration of return-on-investment (ROI) assessment frameworks into predictive maintenance systems can drastically help companies in making financial decisions based on data.

For predictive maintenance, accurate data plays a critical role; however, the systems in the pharmacy industry are often constrained by little data and low measurement quality. Research in the future should focus on the use of new sensor technologies, self-calibrating devices, and data acquisition frameworks that are standardized. Moreover, the use of data fusion techniques and synthetic data production through digital twins and simulation can, especially in the early phases of implementation, alleviate the information scarcity problem. Besides, there has been no shortage of need for research into online real-time sensor networks that will effectively reduce both downtimes and measurement errors.

Predictive algorithms are capable of foreseeing component malfunctions, yet there is always a human operator who plays an essential role in maintenance tasks.

One of the directions for future research is to work on self-aware and self-healing devices that can self-correct. By this, we mean to say that there will be localized intelligence embedded in the components, enabling the assets to autonomously evaluate their condition and decide when to be maintained. Such a development would lead to the industry moving toward fully autonomous adaptive maintenance ecosystems, thus increasing system resilience and reducing the need for human operators.

One of the reasons that predictive maintenance models are not being widely adopted in the industry is that there are still problems related to integration, monitoring, and updating resulting from their use and the nature of the existing operational environment. In this regard, future studies should be oriented towards the deployment of the models complete with automated pipelines, the use of standardized industrial protocols, as well as the establishment of productive interactions between IT professionals and the specialists of the particular domain. On the other hand, adaptive learning methods such as continual learning and federated learning can be seen as shining ways leading to the establishment of the reliability of the models in the long run; thus, researchers will not have to deal with issues like conceptual drift and data contamination.

Reductive maintenance success relies on the cooperation of several disciplines, such as industrial management, computer science, mechanical engineering, and materials science. Next, research will be helped with an open-source repository and social media platform that will facilitate the sharing of model datasets and algorithms. The collaborations will not only encourage newideas but also will make predictive maintenance a vital aspect of Industry 4.0 ecosystems. AI and ML are becoming the main tools in tribology, providing the best modeling and forecasting capabilities for difficult-to-understand friction, lubrication, and wear systems. Constantly monitoring operations, intelligent algorithms can make predictions on material selection, surface engineering, and lubricant formulation that are almost perfect. Moreover, component performance improvement is coupled with predictive maintenance plans that lower the cost of operating the machines and at the same time increase productivity by cutting losses caused by unplanned downtime and failures.

The applications of AI and ML-based techniques in tribology that have been demonstrated to be effective are likely to receive even more support due to the increase in processing power and the facilitation of cross-industry data sharing. Such developments will not only make it possible for the deployment of intelligent algorithms to contribute to the development of reliable, data-based methods for optimizing tribological design and performance but will also reinforce their role. One of the most important directions for future research is the use of AI and ML at the microscopic level, especially in the case of nanoparticles used as solid lubricants and in lubricant acquittal. To achieve the desired tribolayer formation and thus minimize wear under various operating conditions, it is vital to find the best practical particle size and concentration first. The linking of powerful modeling with nanoscale insights will grant researchers the ability to better understand the dynamics of surface topography change during operation and wear progression.

Coating technologies can open up new ways to analyze surfaces and structures across a wide range of engineering domains. They include measurements of surface roughness, evaluations of mechanical integrity properties (such as tensile, compressive, and flexural strengths), and enhancements in solar energy system performance through alterations of optical and thermal properties. Also, coatings make it easier to accurately model how fluids move through complicated networks of channels and transitions (like pipes and elbows). Functional coatings make load mitigation and structural reaction more predictable in precision-engineered systems, including seating systems that are sensitive to ergonomics. This makes the systems last longer, improves the design, and reduces errors in multiphysics simulations [44–61].

4.7 SUMMARY

This chapter wraps up with an assertion that AI and ML have come a long way and are now key forces in predictive maintenance and tribology. They provide cutting-edge technology for industrial system modeling, monitoring, and optimization. The success of these technologies can be seen through their application in the areas of manufacturing surface engineering and condition monitoring, where they have been able to forecast failures, lower downtime, and, consequently, boost efficiency. The improvement in artificial intelligence and the partnership between different fields of study will still bring about significant breakthroughs in the future, even though there are still challenges mainly related to the quality of data, expenses, and the deployment of models. Cutting-edge technologies such as self-aware systems, nanomaterials research, and hybrid AI will be the main drivers, making the whole process of rethinking tribological applications and maintenance plans more efficient, reliable, and sustainable.

REFERENCES

1. Lee, J., Bagheri, B., & Kao, H.-A. (2014). Recent advances and trends of cyber-physical systems and big data analytics in industrial informatics. In *12th IEEE international conference on industrial informatics* (pp. 1–6). Porto Alegre, Brazil: IEEE.
2. Jin, W., Liu, Z., Shi, Z., Jin, C., & Lee, J. (2017). CPS-Enabled worry-free industrial applications. In *2017 Prognostics and system health management conference* (pp. 1–7). Harbin, China: IEEE.
3. Lee, J., Ni, J., Djurdjanovic, D., Qiu, H., & Liao, H. (2006). Intelligent prognostics tools and e-maintenance. *Computers in Industry*, 57(6), 476–489.
4. Sezer, E., Romero, D., Guedea, F., MacChi, M., & Emmanouilidis, C. (2018). An industry 4.0-enabled low-cost predictive maintenance approach for SMEs. In *2018 IEEE International conference on engineering, technology and innovation* (pp. 1–8). Stuttgart, Germany: IEEE.
5. Kiangala, K. S., & Wang, Z. (2018). Initiating predictive maintenance for a conveyor motor in a bottling plant using industry 4.0 concepts. *International Journal of Advanced Manufacturing Technology*, 97(9–12), 3251–3271.
6. Gunes, V., Peter, S., Givargis, T., & Vahid, F. (2014). A survey on concepts, applications, and challenges in cyber-physical systems. *KSII Transactions on Internet and Information Systems*, 8(12), 4242–4268.
7. Wu, D., Jennings, C., Terpenny, J., Kumara, S., & Gao, R. (2017). Cloud-based parallel machine learning for prognostics and health management: A tool wear prediction case study. *Journal of Manufacturing Science and Engineering*, 140(4), 041005.

8. Lee, J., Lapira, E., Bagheri, B., & Kao, H. (2013). Recent advances and trends in predictive manufacturing systems in big data environment. *Manufacturing Letters*, 1(1), 38–41.
9. Lee, J., Bagheri, B., & Kao, H.-A. (2014). Recent advances and trends of cyber-physical systems and big data analytics in industrial informatics. In *12th IEEE international conference on industrial informatics* (pp. 1–6). Porto Alegre, Brazil: IEEE.
10. Muhuri, P. K., Shukla, A. K., & Abraham, A. (2019). Industry 4.0: A bibliometric analysis and detailed overview. *Engineering Applications of Artificial Intelligence*, 78, 218–235.
11. Haarman, M., Mulders, M., & Vassiliadis, C. (2017). *Predictive maintenance 4.0: Predict the unpredictable* (p. 31). PwC and Mainnovation.
12. Mobley, R. (2002). *Introduction to predictive maintenance* (pp. 1–437). Elsevier.
13. Wu, D., Jennings, C., Terpenny, J., & Kumara, S. (2016). Cloud-based machine learning for predictive analytics: Tool wear prediction in milling. In *Proceedings 2016 IEEE international conference on big data* (pp. 2062–2069). IEEE.
14. Terrissa, L. S., Meraghni, S., Bouzidi, Z., & Zerhouni, N. (2016). A new approach of PHM as a service in cloud computing. In *2016 4th IEEE international colloquium on information science and technology* (pp. 610–614). Tangier, Morocco: IEEE.
15. Zerhouni, N., Atamuradov, V., Medjaher, K., Dersin, P., & Lamoureux, B. (2017). Prognostics and health management for maintenance practitioners-review, imple mentation and tools evaluation. *Article in International Journal of Prognostics and Health Management*, 8, 31.
16. Iskandar, J., Moyne, J., Subrahmanyam, K., Hawkins, P., & Armacost, M. (2015). Predictive maintenance in semiconductor manufacturing. In *2015 26th annual SEMI advanced semiconductor manufacturing conference* (pp. 384–389). Institute of Electrical and Electronics Engineers Inc.
17. Nemeth, T., Ansari, F., Sihn, W., Haslhofer, B., & Schindler, A. (2018). PriMa-X: A reference model for realizing prescriptive maintenance and assessing its maturity enhanced by machine learning. *Procedia CIRP*, 72, 1039–1044.
18. Kwon, D., Hodkiewicz, M. R., Fan, J., Shibutani, T., & Pecht, M. G. (2016). IoT-Based prognostics and systems health management for industrial applications. *IEEE Access*, 4, 3659–3670.
19. Zerhouni, N., Atamuradov, V., Medjaher, K., Dersin, P., & Lamoureux, B. (2017). Prognostics and health management for maintenance practitioners-review, imple mentation and tools evaluation. *Article in International Journal of Prognostics and Health Management*, 8, 31.
20. Deutsch, J., & He, D. (2018). Using deep learning-based approach to predict remaining useful life of rotating components. *IEEE Transactions on Systems Man and Cybernetics: Systems*, 48(1), 11–20.
21. Ayad, S., Terrissa, L. S., & Zerhouni, N. (2018). An IoT approach for a smart maintenance. In *2018 International conference on advanced systems and electric technologies* (pp. 210–214). Hammamet, Tunisia: IEEE.
22. Wang, J., Liu, C., Zhu, M., Guo, P., & Hu, Y. (2018). Sensor data based system-level anomaly prediction for smart manufacturing. In *2018 IEEE international congress on big data* (pp. 158–165). San Francisco, CA, USA: IEEE.
23. Lee, J., Ni, J., Singh, J., Jiang, B., Azamfar, M., & Feng, J. (2020). Intelligent maintenance systems and predictive manufacturing. *Journal of Manufacturing Science and Engineering*, 142, 110805.
24. deFaria, H., Costa, J. G. S., & Olivas, J. L. M. (2015). A review of monitoring methods for predictive maintenance of electric power transformers based on dissolved gas analysis. *Renewable and Sustainable Energy Reviews*, 46, 201–209.

25. Levitt, J. (2003). *Complete guide to preventive and predictive maintenance.* Industrial Press Inc. 26. Pech, M., Vrchota, J., & Bednář, J. (2021). Predictive maintenance and intelligent sensors in smart factory: Review. *Sensors*, 21, 1470.
26. Cakir, M., Guvenc, M. A., & Mistikoglu, S. (2020). The experimental application of popular machine learning algorithms on predictive maintenance and the design of IIoT-based condition monitoring system. *Computers & Industrial Engineering,* 151, 106948.
27. Poór, P., Basl, J., & Zenisek, D. (2019, March 28). *Predictive Maintenance 4.0 as next evolution step in industrial maintenance development.* Proceedings of the 2019 International Research Conference on Smart Computing and Systems Engineering (SCSE), Colombo, Sri Lanka, pp. 245–253.
28. Sipos, R., Fradkin, D., Moerchen, F., & Wang, Z. (2014, August 24–27). *Log-based predictive maintenance.* Proceedings of the 20th ACM SIGKDD International Conference on Knowledge Discovery and Data Mining, New York, NY, pp. 1867–1876.
29. Ton, B., Basten, R., Bolte, J., Braaksma, J., Di Bucchianico, A., Van De Calseyde, P., Grooteman, F., Heskes, T., Jansen, N., Teeuw, W., et al. (2020). PrimaVera: Synergising predictive maintenance. *Applied Sciences*, 10, 8348.
30. Vakis, A. I., Yastrebov, V. A., Scheibert, J., Nicola, L., Dini, D., Minfray, C., Almqvist, A., Paggi, M., Lee, S., & Limbert, G. (2018). Modeling and simulation in tribology across scales: An overview. *Tribology International*, 125, 169–199.
31. Burge, P. (2013). SKF launches SKF Insight, groundbreaking intelligent bearing technology. Retrieved 2022, from http://news.cision.com/skf-uk/r/skf-launches-skf-insight-groundbreaking-intelligent-bearing-technology%2Cc9401490.
32. igus. (2022). *Smart bearing technology.* https://www.igus.co.uk/info/smart-bearings.
33. Macmillan, D., Smith, E. H., Sherrington, I., & Johns-Rahnejat, P. (2023). *Development of a novel test apparatus for the evaluation of an active geometry tilting-pad thrust bearing and its comparison with a conventional passive type.* Proceedings LUBMAT 2023.
34. Ulbrich, H., & Althaus, J. (1989). Actuator design for rotor control.
35. Santos, I. F. (2010). *Mechatronics applied to machine elements with focus on active control of bearing, shaft and blade dynamics.* DTU Mechanical Engineering. https://backend.orbit.dtu.dk/ws/portalfiles/portal/5561415/santos_final.pdf.
36. Wirsching, S., Marian, M., Bartz, M., et al. (2021). Geometrical optimization of the EHL roller face/rib contact for energy efficiency in tapered roller bearings. *Lubricants*, 9, 67. https://doi.org/10.3390/lubricants9070067
37. Roy, B., & Dey, S. (2021). Machine learning-based performance analysis of two-axial-groove hydrodynamic journal bearings. *Proceedings of the Institution of Mechanical Engineers, Part J: Journal of Engineering Tribology, 235*, 2211–2224. https://doi.org/10.1177/1350650121992895
38. Schwarz, S., Grillenberger, H., Tremmel, S., & Wartzack, S. (2021). Prediction of rolling bearing cage dynamics using dynamic simulations and machine learning algorithms. *Tribology Transactions.* https://doi.org/10.1080/10402004.2021.1934618
39. SKF. (n.d.). *Condition monitoring systems.* https://www.skf.com/uk/products/condition-monitoring-systems.
40. Paturi, U. M. R., Palakurthy, S. T., & Reddy, N. S. (2023). The role of machine learning in tribology: A systematic review. *Archives of Computational Methods in Engineering*, 30, 1345.
41. Alambeigi, F., Khadem, S. M., Khorsand, H., & Mirza Seied Hasan, E. (2016). [PDF] from researchgate.netA comparison of performance of artificial intelligence methods in prediction of dry sliding wear behavior. *The International Journal of Advanced Manufacturing Technology*, 84, 1981.

42. King, K. (2020). *Coefficient of determination and root-mean-square error (RMSE)*. https://www.kristakingmath.com/blog/coefficient-of-determination-and-root-mean-square-error.
43. Pimenov, D. Y., Da Silva, L. R. R., Ercetin, A., Der, O., Mikolajczyk, T., & Giasin, K. (2024). State-of-the-art review of applications of image processing techniques for tool condition monitoring on conventional machining processes. *The International Journal of Advanced Manufacturing Technology*, 130, 57.
44. Verma, P. K., Vasudev, H., & Kumar, V. (2025). Dental implant materials and designs and failure during implant placement and loading: A review. In *Advanced manufacturing processes* (pp. 138–150). Springer Nature.
45. Verma, P. K., Vasudev, H., & Kumar, V. (2025). Bioactivity and corrosion performance of low-velocity oxy-fuel-sprayed hydroxyapatite-based coatings on a surgical-grade stainless steel. *Transactions of the Indian Institute of Metals,* 78(8), 185.
46. Verma, P. K., Vasudev, H., & Kumar, V. (2025). Effect of electro-corrosion behavior of different grades of stainless steel used in bio-implants: A review. In *Advanced manufacturing processes* (pp. 269–283). CRC Press.
47. Verma, P. K., & Vasudev, H. (2025). Importance of hydroxyapatite coatings in improving new bone osteogenesis: A review. In *Advanced manufacturing processes* (pp. 284–299). CRC Press.
48. Dutta, V., Thakur, L., Singh, B., & Vasudev, H. (2022). A study of erosion–corrosion behaviour of friction stir-processed chromium-reinforced NiAl bronze composite. *Materials,* 15(15), 5401.
49. Singh, M., Vasudev, H., & Kumar, R. (2021). Corrosion and tribological behaviour of BN thin films deposited using magnetron sputtering. *International Journal of Surface Engineering and Interdisciplinary Materials Science,* 9(2), 24–39.
50. Sharanabasava, H., Raviprakash, M., Prasad, C. D., Ramesh, M. R., Phanibhushana, M. V., Vasudev, H., & Kumar, S. (2023). Microstructure, mechanical and wear properties of SiC and Mo reinforced NiCr microwave cladding. *Advances in Materials and Processing Technologies,* 10, 3620–3633.
51. Singh, M., Vasudev, H., & Singh, M. (2022). Surface protection of SS-316L with boron nitride based thin films using radio frequency magnetron sputtering technique. *Journal of Electrochemical Science and Engineering*, 12(5), 851–863.
52. Prashar, G., Vasudev, H., Thakur, L., & Bansal, A. (2023). Performance of thermally sprayed hydroxyapatite coatings for biomedical implants: A comprehensive review. *Surface Review and Letters*, 30(1), 2241001.
53. Vasudev, H. (2020). Wear characteristics of Ni-WC powder deposited by using a microwave route on mild steel: Microwave cladding of Ni-WC. *International Journal of Surface Engineering and Interdisciplinary Materials Science*, 8(1), 44–54.
54. Parkash, J., Saggu, H. S., & Vasudev, H. (2022). A short review on the performance of high velocity oxy-fuel coatings in boiler steel applications. *Materials Today: Proceedings*, 50, 1442–1446.
55. Singh, P., Vasudev, H., & Bansal, A. (2024). Effect of post-heat treatment on the microstructural, mechanical, and bioactivity behavior of the microwave-assisted alumina-reinforced hydroxyapatite cladding. *Proceedings of the Institution of Mechanical Engineers, Part E: Journal of Process Mechanical Engineering*, 238(3), 1060–1071.
56. Verma, P. K., Vasudev, H., Singh, G., & Singh, S. (2024). A review on the microstructural and biomedical properties of implants manufactured using additive manufacturing. *Journal of the Institution of Engineers (India): Series D*, 106, 1–15.
57. Singh, S. P., Verma, P. K., Singh, S., Kapoor, M., & Balu, R. (2024). Microstructural characterization and electrochemical behaviour of HVOF sprayed WC-Co-Cr coatings on 316L boiler steel stainless steel. *Results in Surfaces and Interfaces*, 16, 100248.

58. Verma, P. K., Minhas, A. S., Singh, P., & Kumar, S. (2024). Slurry erosion behaviour of thermal sprayed Al_2O_3 and Cr_2O_3 coatings for turbine steels. AIP Conference Proceedings. AIP Publishing LLC, Vol. 2986, No. 1, p.020012.
59. Verma, P. K., Kumar, V., & Vasudev, H. (2024). Bioactivity and corrosion analysis of thermally sprayed hydroxyapatite based coatings. *Journal of Electrochemical Science and Engineering*, 14(4), 419–439.
60. Verma, P. K., Vasudev, H., & Kumar, V. (2025). Biocompatibility and corrosion performance of HA-based LVOF sprayed and post-annealed coatings on SS 254 surgical-grade bio-implant material. *Ceramics International*, 51(10), 13207–13225.
61. Verma, P. K., Vasudev, H., & Singh, G. (2025). Improving the corrosion performance of SS 304 stainless steel using HVOF-sprayed WC-Co-Cr coatings in chloride environments. *Journal of Bio- and Tribo-Corrosion*, 11(4), 1–13.

5 Enhancing Tribological Performance with Artificial Intelligence and Machine Learning Techniques

Gaurav Prashar and Hitesh Vasudev

5.1 INTRODUCTION

Artificial Intelligence (AI) is scientific and engineering computing driven by human intelligence to achieve automated learning through various techniques initiated by researchers. Inherently, AI is transformed from natural intelligence [1]. Between 1950 and 2020, several approaches have been implemented so far to achieve required error-free and successful data collection to drive AI in the manufacturing industry [2]. As global competition is becoming increasingly considerable, the adoption rate for modern industry levels is also changing from Industry 4.0 to Industry 5.0 to enhance efficiency and provide cost-effective solutions to optimize production [3]. Human–machine collaboration is taking place to successfully implement AI in industry [4]. Since the introduction of real-life implementation of AI, it has always faced issues and challenges. These hurdles are always related to, but not limited to, data collection, data quality, and the selection of AI models to interpret data in an organized manner [3].

If we look at the past, humans have revolutionized in terms of skill development and self-learning while hunting in the Stone Age. In that era, humans transformed their ordinary way of hunting into a sophisticated one by learning strategies according to the behavior of animals. Humans developed further protection arrangements to enhance the potential of hunting in several ways, such as by intercommunication, using shields, and problem-solving techniques. Consequently, humans developed agricultural methods, animal farming, and tools to conduct these practices. In this self-learning process, their curiosity to learn about the environment and self-betterment developed human intelligence [5]. To address these challenges and with further development of human intelligence, several approaches have been introduced to date

DOI: 10.1201/9781003741152-5

by a number of researchers between 1950 and 2020. The overall development of AI consists of three different stages: AI infancy, and model-based and data-driven approaches [6–12].

5.1.1 History and Architecture of Artificial Intelligence (AI)

If we talk about AI's first contribution, the development goes back to 1958 when the design of programming languages developed [13]. In this era, symbols were used to interpret rules and calculations [14]. This contribution is regarded as the very first in AI. Before Rosenblatt in 1957, neurophysiologists had introduced basic calculations to visualize behavior [15]. As per prescribed, regulated, human-designed rules and knowledge models, it is often regarded as the first AI input–output model machine, but it had limited applicability and potential. So this model remained unused further [16]. The development of AI was resumed in the 1980s when computers were introduced to store commands such as "if" and "then" [17]. This modified model and integration of computers in AI paved the way to diagnose several medical issues related to blood. Later, using this approach, more issues were tackled in the areas of locomotive failures [18] and to find geological deposits [19]. In 1986, backpropagation was developed to optimize any available AI model [20], resulting in the breakthrough of concurrent neural networks, such as convolutional neural networks (CNNs) and recurrent neural networks (RNNs) [21, 22].

Later, it was shown by the universal approximation theorem that these functions can be fitted to any function [23]. Notably, two types of techniques were revolutionized: model based and data-driven based [24]. In the 1990s, kernel methods were developed. In this regard, support vector machines (SVMs) thrived [25]. SVMs possess small datasets. This requirement and applicability made SVMs more usable than neural networks [26]. Mitchell in 1997 provided a definition of machine learning (ML), describing that neural networks can learn from data without the need for external commands [27]. These ML processes can be categorized into three approaches: supervised learning, unsupervised learning, and reinforcement learning. In supervised learning, the goal is to minimize the difference between predicted and actual outcomes. The outcomes are derived from the input data. This data helps the ML model to learn from outcomes. Whereas in unsupervised learning, the ML model is trained without any response or outcome. In this learning approach, the main goal is to recognize patterns, clusters, and overall relationships within the data itself. Finally, in reinforcement learning models, the learning process typically holds to the actions taken by a particular data-driven subset. The model trains itself by receiving feedback from previous actions and making corrections where required. There is no direct involvement of direct labeled instructions.

5.1.2 Expansion of AI in Tribology

In the last decade, significant testing of AI has been achieved to enhance digital transformation. Tribological materials play a critical role in reducing friction, wear, and lubrication-related failures in multiple industries, such as the aerospace, automotive, and biomedical sectors. The selection and development of these materials have historically depended on experimental testing and computational modeling. These methods have significantly advanced the field; however, they tend to be time-consuming, costly to implement, and limited in their capacity to capture complex, nonlinear interactions among multiple variables, including load, sliding speed, material composition, temperature, and surface conditions. Recent advancements in ML provide significant alternatives by facilitating the analysis of extensive datasets to uncover hidden patterns and accurately predict material properties.

5.1.3 Large Datasets, Computational Power, and Deep Learning

During the stagnation period in the development of AI, large datasets often created problems in result findings due to complexity in big data. This challenge is overcome by the introduction of deep neural networks and computational power to solve problems related to datasets. In this regard, graphic processing units (GPUs) and cloud computing were invented to meet the requirements [28]. Scientists have tested the potential of deep learning neural networks to solve datasets. They examined and found that this model exploits the hierarchical structure of data under consideration. They have also seen that no excessive examination of data is required during the deep learning process in ML, as was required in SVM [29]. Furthermore, empirical studies showed that errors associated with a lack of theoretical knowledge are being surpassed by the excellence of data handling by the deep learning approach. Its superior performance in handling and managing datasets in image processing far outperforms previous models [30, 31]. In this period, deep learning was gaining traction in ML with the help of deep neural networks, which require almost no data input sorting like SVM. The application of ML in general materials design is well established; however, its use in predicting tribological behavior, specifically in relation to friction and wear characteristics, is still relatively underexplored. This review focuses on ML models that predict tribological properties.

5.1.4 ML in Tribological Material Prediction

Figure 5.1 outlines the method employed for predicting TM properties utilizing an ML approach. Experimental measurement by performing actual experimentation and computer modeling are the two prime approaches for studying material characteristics [32]. Both methods are fundamentally correct but may present certain difficulties in completely highlighting the intricate relationships among TM properties and the associated factors [33]. Therefore, if the results from modeling or experimental testing are not good, resources like effort, time, and cost associated may become unproductive [34–36].

A notable issue is that it can be extremely challenging to know all TM properties using only computational or experimental approaches, specifically when dealing with composite materials or a large number of parameters [37]. Therefore, it becomes relatively important to develop and design better prediction models that predict TM properties accurately while reducing computational costs and time. The

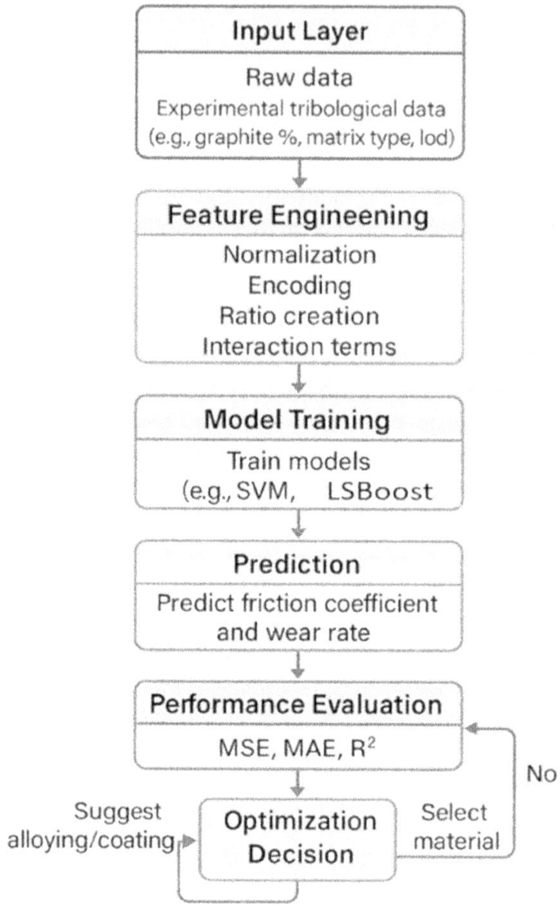

FIGURE 5.1 ML approach for the forecasting and optimization of TM [41].

answer is the development and design of ML algorithms that can pin down patterns in large datasets, permitting faster and better forecasting of TM properties [38]. The adaptability of ML and its capability to establish across complex systems have developed a new paradigm in TM research [39].

A standard ML system is depicted in Figure 5.2. The prediction system uses the trained models to recommend possible material structures obtained from the given inputs [40, 41].

5.1.5 Tribology and ML

To design better and reliable tribo-engineering components, it is important to have a thorough understanding of the TM behavior. TM datasets act as vital input for ML algorithms, allowing them to utilize existing knowledge to further enhance the

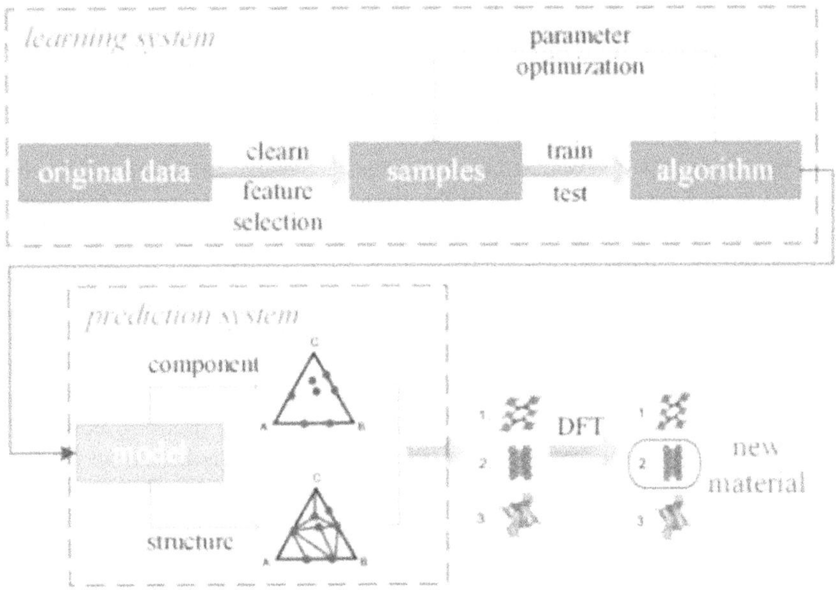

FIGURE 5.2 Standard ML approach in discovering novel materials [41].

components' performance [42]. The properties of TM are governed by a range of factors, such as the environment and operating conditions. Knowing the impact of these variables on tribological performance is further important for TM properties [43].

ML techniques/algorithms may deliver a viable answer to these associated factors [44]. Hence, the use of novel algorithms in the field of tribology allows for a more knowledgeable comprehension of tribological phenomena. In tribology, ML serves different roles, for example, condition monitoring of parts, and predicting wear rates (WRs) and friction coefficients (FCs) [45].

5.1.6 ML Techniques

Neural networks (NNs), feed-forward multilayered perceptron (FFMP), and the radial basis function algorithm (RBFA) have been employed to forecast WR and determine FC with better accuracy.

Decision trees and their associated variants, commonly known as random forest (RF) and bagging, are accepted by scholars for their interpretability and potential to govern mixed data types. In the field of tribology, they are commonly used to assess WR across different operational circumstances. One major disadvantage associated with them is their tendency to overfit, particularly when data is less. Support vector machines deliver high accuracy and display effective results/outcomes in high-dimensional spaces, rendering them suitable for estimating FC across different loads or temperature ranges. Their primary disadvantages are reduced interpretability and high sensitivity to the selection of kernel parameters.

Artificial neural networks (ANNs) are widely utilized in the TM research domain that involves nonlinear, multivariable relationships, including the prediction of FC and WR in coatings. However, they may function as black box models, which limits their interpretability. Another ML ensemble method, least squares boosting, depends on the gradient boosting framework, which integrates several weak learners to improve prediction performance. Their effectiveness is required where high accuracy is important, particularly in the simultaneous prediction of FC and WR. However, these systems need increased computational resources and may further diminish interpretability.

5.2 STUDIES RELATED TO WEAR PREDICTION

Ten sets of different ML models were employed by Sivaraman et al. [46] for predicting WR and phase formation in high entropy alloy coating (HEAC). Four of the models were used for phase prediction (PP). The outcomes indicate that XGBoost has the highest level of predictive effectiveness for WR, obtaining an R^2 value of 0.98. The RF model demonstrated the highest accuracy of 98.5% in PP for novel HEAC, as depicted in Figure 5.3. The outcomes underscore the capability of ML techniques in enhancing TM design and optimizing coatings.

The friction behavior of FeCoCrNiAlN HEAC was analyzed using seven ML regression models. The XGBoost regressor and GBRT exhibited the highest predictive accuracy, achieving R^2 values of 0.86 and 0.81, respectively [47]. The identification of phases in CoCrFeNiMn HEAC was estimated using a NN model, achieving an accuracy of 94% [48].

The experimental work by Suryawanshi et al. [49] conducted the development and optimization of a three-layer ANN model for evaluating the WR and FC of deposited

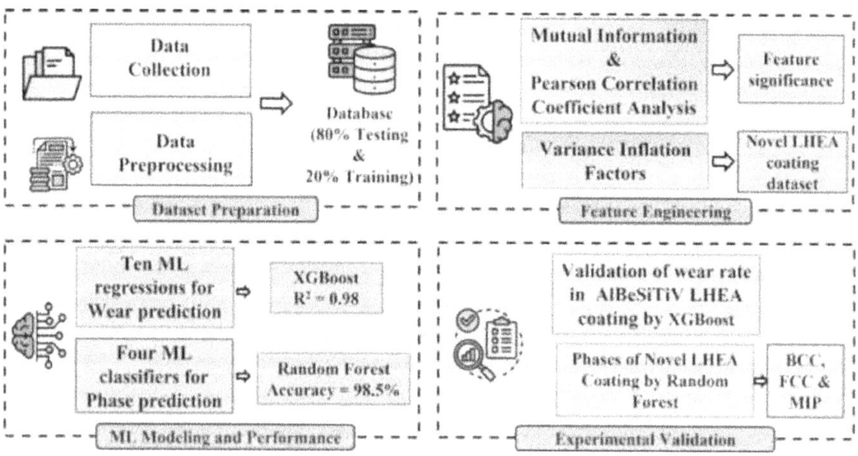

FIGURE 5.3 Prediction of WR and PP in HEAC utilizing different ML models [46]. (Creative Common license.)

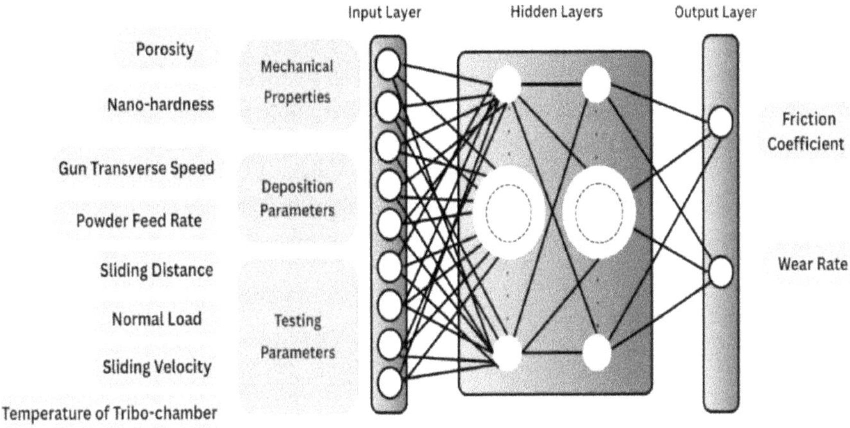

FIGURE 5.4 Three-layer ANN structure showing the relationship between the input and output layer [49]. (Creative common license.)

coatings (Figure 5.4). The low mean relative errors of 0.78% for FC and 0.93% for WR, along with the high (R^2 = 0.99144 for FC and R^2 = 0.99424 for WR), clearly indicate the model's effectiveness.

Wear is the most common problem encountered in industrial applications [50–57]. WC–Co coatings as confirmed in the literature, exhibit higher wear resistance and better mechanical durability; however, precisely forecasting their tribological performance for vital parts remains an issue for researchers due to the complexity of associated wear mechanisms and material heterogeneity. Standard experimental test techniques are often associated with economic considerations, making it challenging for scientists/professionals to develop an accurate predictive framework. Furthermore, current experimental models exhibit a constrained capacity for varying operating conditions during wear tests. The identified merits indicate a need for the development and design of more advanced algorithms/computational techniques, such as ML, which can leverage experimental data to the next level for generating effective, flexible, and better models of prediction for the tribological performance of materials.

Shabana and co-workers [58] calculated the WR and FC for WC–Co coating. Confirmation experiments on coatings further established that the support vector regression model is the most appropriate model for predicting WR, whereas Gaussian process regression showcases superior performance in predicting FC. The findings highlight that ML can decrease the number of effort-consuming experimental trials, leading to optimal TM selection and improved performance for applications in important industrial manufacturing sectors. The results demonstrate that ML is a dependable method for predicting WR and FC, thereby facilitating novel insights in data-driven intelligent tribology.

Coating technologies can help engineers in many fields look at surfaces and structures in new ways. They include tests of surface roughness, tests of mechanical integrity features (including tensile, compressive, and flexural strengths), and improvements in the performance of solar energy systems by changing their optical and thermal properties. Coatings also help you effectively describe how fluids move through networks of channels and transitions that are hard to understand (like pipes and elbows). Functional coatings make it easier to predict how loads would affect the structure and how it will behave in precision-engineered systems, such as seating systems that are sensitive to ergonomics. This makes the systems last longer, makes the design better, and cuts down on mistakes in multiphysics simulations [59–76].

5.3 CONCLUSION

This short review chapter stresses the capabilities of ML in transforming the discovery and optimization processes of TM. The integration of ML with tribological assessments has led to the development of novel insights aimed at enhancing TM selection for a variety of high-performance parts. The outcome of the review study reveals that ML models/algorithms can accurately predict TM performance, thereby assisting in the TM selection that meets tribological standards. Future work in the tribological field should stress materials like ceramics and polymers, incorporating advanced approaches such as explainable AI to enhance interpretability. The integration of TM databases and the implementation of automated ML pipelines can remarkably enhance the efficiency of the discovery path. This review highlights that the integration of ML within the tribological field has significant potential for reducing development timelines, improving performance, and identifying novel material solutions suited to different industrial requirements.

REFERENCES

1. High, P. (2017, October 30). *Carnegie Mellon dean of computer science on the future of AI*. Forbes.
2. Gao, R. X., Krüger, J., Merklein, M., Möhring, H. C., & Váncza, J. (2024). Artificial intelligence in manufacturing: State of the art, perspectives, and future directions. *CIRP Annals, 73*, 723–749.
3. Akhtar, Z. B. (2024). Artificial intelligence (AI) within manufacturing: An investigative exploration for opportunities, challenges, future directions. *Metaverse, 5*(2), 27–31.
4. Archana, T., & Stephen, R. K. (2024). The future of artificial intelligence in manufacturing industries. In *Industry applications of thrust manufacturing: Convergence with real-time data and AI* (pp. 98–117). IGI Global.
5. Sagan, C., & Druyan, A. (1993). *Shadows of forgotten ancestors: A search for who we are*. Random House.
6. Bellman, R. (1954). The theory of dynamic programming. *Bulletin of the American Mathematical Society, 60*(6), 503–515.
7. Breiman, L. (2001). Random forests. *Machine Learning, 45*, 5–32.
8. Morgan, J. N., & Sonquist, J. A. (1963). Problems in the analysis of survey data, and a proposal. *Journal of the American Statistical Association, 58*(302), 415–434.
9. OpenAI. (2024). *Introducing ChatGPT.*

10. Shortliffe, E. H. (1976). MYCIN: A rule-based computer program for advising physicians regarding antimicrobial therapy selection. In *Proceedings of the AFIPS '76 National Computer Conference* (pp. 739–739). https://doi.org/10.1145/1408800.1408906.
11. Wang, L. (2019). From intelligence science to intelligent manufacturing. *Engineering, 5*(4), 615–618. https://doi.org/10.1016/j.eng.2019.04.014.
12. Zadeh, L. A. (1965). Fuzzy sets. *Information and Control, 8*(3), 338–353.
13. McCarthy, J. (1978). *History of LISP.* Academic Press.
14. Wen, B., Lian, W., Bekris, K., & Schaal, S. (2022). Catgrasp: Learning category-level task-relevant grasping in clutter from simulation. In *2022 IEEE International Conference on Robotics and Automation (ICRA)* (pp. 6401–6408). IEEE. https://doi.org/10.1109/ICRA46639.2022.9812273.
15. McCulloch, W. S., & Pitts, W. (1943). A logical calculus of the ideas immanent in nervous activity. *The Bulletin of Mathematical Biophysics, 5*(4), 115–133.
16. Minsky, M. L., & Papert, S. A. (1988). *Perceptrons* (Expanded ed.). MIT Press.
17. Tan, H. (2017). A brief history and technical review of the expert system research. *IOP Conference Series: Materials Science and Engineering, 242*, 012111. https://doi.org/10.1088/1757-899X/242/1/012111.
18. Bonissone, P. P., & Johnson, H. E. (1984). *DELTA: An expert system for diesel electric locomotive repair.* General Electric Corporate Research and Development.
19. Gaschnig, J. (1982). Prospector: An expert system for mineral exploration. In D. Michie (Ed.), *Introductory readings in expert systems* (pp. 47–64). Gordon and Breach Science Publishers.
20. Rumelhart, D. E., Hinton, G. E., & Williams, R. J. (1986). Learning representations by back-propagating errors. *Nature, 323*(6088), 533–536. https://doi.org/10.1038/323533a0.
21. Hochreiter, S., & Schmidhuber, J. (1997). Long short-term memory. *Neural Computation, 9*(8), 1735–1780.
22. LeCun, Y., Boser, B., Denker, J., Henderson, D., Howard, R., Hubbard, W., & Jackel, L. (1989). Handwritten digit recognition with a back-propagation network. In D. S. Touretzky (Ed.), *Advances in neural information processing systems* (Vol. 2, pp. 396–404). Morgan Kaufmann.
23. Hornik, K., Stinchcombe, M., & White, H. (1989). Multilayer feedforward networks are universal approximators. *Neural Networks, 2*(5), 359–366.
24. Holland, J. H. (1992). *Adaptation in natural and artificial systems: An introductory analysis with applications to biology, control, and artificial intelligence.* MIT Press.
25. Cortes, C., & Vapnik, V. (1995). Support-vector networks. *Machine Learning, 20*(3), 273–297.
26. Hofmann, T., Schölkopf, B., & Smola, A. J. (2008). Kernel methods in machine learning. *The Annals of Statistics, 36*(3), 1171–1220.
27. Mitchell, T. M. (1997). *Machine learning.* McGraw-Hill.
28. Krizhevsky, A., Sutskever, I., & Hinton, G. E. (2012). ImageNet classification with deep convolutional neural networks. In F. Pereira, C. J. C. Burges, L. Bottou, & K. Q. Weinberger (Eds.), *Advances in neural information processing systems* (Vol. 25, pp. 1097–1105). Curran Associates, Inc.
29. LeCun, Y., Bengio, Y., & Hinton, G. (2015). Deep learning. *Nature, 521*(7553), 436–444. https://doi.org/10.1038/nature14539.
30. Farabet, C., Couprie, C., Najman, L., & LeCun, Y. (2012). Learning hierarchical features for scene labeling. *IEEE Transactions on Pattern Analysis and Machine Intelligence, 35*(8), 1915–1929.

31. Hinton, G., Deng, L., Yu, D., Dahl, G. E., Mohamed, A. R., Jaitly, N., ... & Kingsbury, B. (2012). Deep neural networks for acoustic modeling in speech recognition: The shared views of four research groups. *IEEE Signal Processing Magazine, 29*(6), 82–97.
32. Bishara, D., Xie, Y., Liu, W. K., & Li, S. (2023). A state-of-the-art review on machine learning-based multiscale modeling, simulation, homogenization and design of materials. *Archives of Computational Methods in Engineering, 30*(1), 191–222.
33. Sharma, S., Joshi, S. S., Pantawane, M. V., Radhakrishnan, M., Mazumder, S., & Dahotre, N. B. (2023). Multiphysics multi-scale computational framework for linking process–structure–property relationships in metal additive manufacturing: A critical review. *International Materials Reviews, 68*(7), 943–1009.
34. Zamani, V., Taghaddos, H., & Gholipour, Y. (2023). Simulation-based decision support system for earthmoving operations using computer vision. *Engineering Applications of Artificial Intelligence, 124*, 106564. https://doi.org/10.1016/j.engappai.2023.106564.
35. Liskiewicz, T., Sherrington, I., Khan, T., & Liu, Y. (2023, November 1). Advances in sensing for real-time monitoring of tribological parameters. *Tribology International, 189*, 108965. https://doi.org/10.1016/j.triboint.2023.108965.
36. Leary, M. (2011). Materials selection and substitution using aluminium alloys. In R. Lumley (Ed.), *Fundamentals of aluminium metallurgy* (pp. 784–827). Woodhead Publishing.
37. Suki, B. (2022). Selected examples of tissue-level collagen suprastructures: Tendon, bone, and skin. In *Structure and function of the extracellular matrix* (pp. 113–142). Academic Press.
38. Mobarak, M. H., Mimona, M. A., Islam, M. A., Hossain, N., Zohura, F. T., Imtiaz, I., & Rimon, M. I. H. (2023). Scope of machine learning in materials research—A review. *Applied Surface Science Advances, 18*, 100523. https://doi.org/10.1016/j.apsadv.2023.100523.
39. Fang, J., Xie, M., He, X., Zhang, J., Hu, J., Chen, Y., & Yang, Y. (2022). Machine learning accelerates the materials discovery. *Materials Today Communications, 33*, 104900. https://doi.org/10.1016/j.mtcomm.2022.104900.
40. Sarker, I. H. (2021). Machine learning: Algorithms, real-world applications and research directions. *SN Computer Science, 2*(3), 160. https://doi.org/10.1007/s42979-021-00592-x.
41. Zahoor, A., Rahman, A., & Mursaleen, M. (2025). Accelerating tribological material discovery: A machine learning framework with predictive insights. *Tribology International, 211*, 110862.
42. Baş, H., & Karabacak, Y. E. (2023). Machine learning-based prediction of friction torque and friction coefficient in statically loaded radial journal bearings. *Tribology International, 186*, 108592. https://doi.org/10.1016/j.triboint.2023.108592.
43. Chaudhary, V. (2023). Tribology of polymer films and coatings. In S. C. George, J. T. Haponiuk, S. Thomas, R. Reghunath, & S. P. S. (Eds.), *Tribology of polymers, polymer composites, and polymer nanocomposites* (pp. 335–355). Elsevier.
44. Harris, M. D., Berkebile, S. P., Murthy, N. K., & Voevodin, A. A. (2020). In-situ instrumentation to observe transient tribological effects in high sliding speed unlubricated contacts. *Wear, 442–443*, 203111. https://doi.org/10.1016/j.wear.2019.203111.
45. Soni, A., Das, P. K., Yusuf, M., & Ridha, S. (2022). Tribological behavior of particulates reinforced sustainable composites: Effect of composition, load, and sliding speed. *Sustainable Chemistry and Pharmacy, 29*, 100748. https://doi.org/10.1016/j.scp.2022.100748.
46. Sivaraman, S., Radhika, N., & Khan, M. A. (2025). Machine learning-driven prediction of wear rate and phase formation in high entropy alloy coatings for enhanced durability and performance. *IEEE Access, 13*, 33956–33975.

47. Jia, B., Wan, Q., Yan, L., Luo, Y., Wei, Q., Niu, C., Yang, B., Li, S., & Meng, L. (2024). Tribological properties and machine learning prediction of FeCoCrNiAlN high entropy coatings. *Surface and Coatings Technology, 477*, 130341.
48. Gao, S., Gao, Z., & Zhao, F. (2023). Machine learning and visualization assisted solid solution strengthening phase prediction of high entropy alloys. *Materials Today Communications, 35*, 105894.
49. Suryawanshi, S., Bhosale, D. G., Vasudev, H., Rukhande, S., & Panwar, V. (2025). Artificial neural network-based prediction of tribological performance of coatings fabricated by thermal spray techniques for nuclear energy applications. *Franklin Open*, 100364.
50. Krishnamoorthy, R. R., Rozani, N., & Marius, D. (2025). Mechanical and stability testing of aerospace materials. In *Aerospace materials* (pp. 75–101). Elsevier.
51. Li, Y., Xie, G., Sun, X., Jingcao, C., Wu, X., Zhang, Y., Yu, C., & Du, P. (2025). A review on fretting wear/fatigue behavior, protective measures, and application examples of typical alloy materials. *Proceedings of the Institution of Mechanical Engineers, Part J: Journal of Engineering Tribology*, 13506501241313249.
52. Wu, Q., Long, W., Zhang, L., & Zhao, H. (2024). A review on ceramic coatings prepared by laser cladding technology. *Optics & Laser Technology, 176*, 110993.
53. Prasad, C. D., Tiwari, A., Suryawanshi, S. R., Dileep, B. P., Gowda, A. C., Masum, H., Dutt, K. M., Kumar, S. P., & Bavan, S. (2025). Overview of thermal spray coating on additive manufacturing. *Progress in Additive Manufacturing, 10*(8), 4327–4347.
54. Psarommatis, F., & May, G. (2025). A cost–benefit model for sustainable product reuse and repurposing in circular remanufacturing. *Sustainability, 17*(1), 245.
55. Yazdi, M. (2024). Maintenance strategies and optimization techniques. In *Advances in computational mathematics for industrial system reliability and maintainability* (pp. 43–58). Springer Nature Switzerland.
56. Chen, X., Chen, Z., Wei, T., Zhan, D., Huang, J., & Guo, Z. (2025). Synergistic enhancement of wear resistance via CrN coating and layered double hydroxides systems. *Tribology International, 207*, 110540.
57. Wang, H., Wang, C., Dang, J., Gao, P., An, Q., Zheng, T., Deng, Y., Ming, W., Chen, M., Wang, T., & Yang, L. (2025). Frication behaviors, wear and failure mechanisms of CBN tool and machined surface morphology in high-speed turning of high strength refractory alloy cast steel. *Journal of Manufacturing Processes, 138*, 186–202.
58. Shabana, S., Jatavallabhula, J. K., Kumar, R. N., Chinthalapudi, R., & Pappula, B. (2025). Towards intelligent tribology: Predicting wear and friction in WC–Co coatings with machine learning. *Journal of Bio- and Tribo-Corrosion, 11*(3), 88.
59. Verma, P. K., Vasudev, H., & Kumar, V. (2025). Dental implant materials and designs and failure during implant placement and loading: A review. In H. Vasudev, C. Prakash, M. Khalid, & KesariK. (Eds.), *Advanced manufacturing processes* (pp. 138–150). CRC Press. https://doi.org/10.1201/9781003476238
60. Verma, P. K., Vasudev, H., & Kumar, V. (2025). Bioactivity and corrosion performance of low-velocity oxy-fuel-sprayed hydroxyapatite-based coatings on a surgical-grade stainless steel. *Transactions of the Indian Institute of Metals, 78*(8), 185.
61. Verma, P. K., Vasudev, H., & Kumar, V. (2025). Effect of electro-corrosion behavior of different grades of stainless steel used in bio-implants: A review. In H. Vasudev, C. Prakash, M. Khalid, & KesariK. (Eds.), *Advanced Manufacturing Processes* (pp. 269–283). CRC Press.
62. Verma, P. K., & Vasudev, H. (2025). Importance of hydroxyapatite coatings in improving new bone osteogenesis: A review. In *Advanced Manufacturing Processes* (pp. 284–299). CRC Press.

63. Dutta, V., Thakur, L., Singh, B., & Vasudev, H. (2022). A study of erosion–corrosion behaviour of friction stir-processed chromium-reinforced NiAl bronze composite. *Materials, 15*(15), 5401.
64. Singh, M., Vasudev, H., & Kumar, R. (2021). Corrosion and tribological behaviour of BN thin films deposited using magnetron sputtering. *International Journal of Surface Engineering and Interdisciplinary Materials Science, 9*(2), 24–39.
65. Sharanabasava, H., Raviprakash, M., Prasad, C. D., Ramesh, M. R., Phanibhushana, M. V., Vasudev, H., & Kumar, S. (2023). Microstructure, mechanical and wear properties of SiC and Mo reinforced NiCr microwave cladding. *Advances in Materials and Processing Technologies, 10*, 3620–3633.
66. Singh, M., Vasudev, H., & Singh, M. (2022). Surface protection of SS-316L with boron nitride based thin films using radio frequency magnetron sputtering technique. *Journal of Electrochemical Science and Engineering, 12*(5), 851–863.
67. Prashar, G., Vasudev, H., Thakur, L., & Bansal, A. (2023). Performance of thermally sprayed hydroxyapatite coatings for biomedical implants: A comprehensive review. *Surface Review and Letters, 30*(1), 2241001.
68. Verma, P. K., Minhas, A. S., Singh, P., & Kumar, S. (2024). Slurry erosion behaviour of thermal sprayed Al_2O_3 and Cr_2O_3 coatings for turbine steels. In *AIP Conference Proceedings* (Vol. 2986, No. 1, p. 020012). AIP Publishing LLC.
69. Verma, P. K., Kumar, V., & Vasudev, H. (2024). Bioactivity and corrosion analysis of thermally sprayed hydroxyapatite based coatings. *Journal of Electrochemical Science and Engineering, 14*(4), 419–439.
70. Verma, P. K., Vasudev, H., & Kumar, V. (2025). Biocompatibility and corrosion performance of HA-based LVOF sprayed and post-annealed coatings on SS 254 surgical-grade bio-implant material. *Ceramics International, 51*(10), 13207–13225.
71. Verma, P. K., Vasudev, H., & Singh, G. (2025). Improving the corrosion performance of SS 304 stainless steel using HVOF-sprayed WC–Co–Cr coatings in chloride environments. *Journal of Bio- and Tribo-Corrosion, 11*(4), 1–13.
72. Vasudev, H. (2020). Wear characteristics of Ni–WC powder deposited by using a microwave route on mild steel: Microwave cladding of Ni–WC. *International Journal of Surface Engineering and Interdisciplinary Materials Science, 8*(1), 44–54.
73. Parkash, J., Saggu, H. S., & Vasudev, H. (2022). A short review on the performance of high velocity oxy-fuel coatings in boiler steel applications. *Materials Today: Proceedings, 50*, 1442–1446.
74. Singh, P., Vasudev, H., & Bansal, A. (2024). Effect of post-heat treatment on the microstructural, mechanical, and bioactivity behavior of microwave-assisted alumina-reinforced hydroxyapatite cladding. *Proceedings of the Institution of Mechanical Engineers, Part E: Journal of Process Mechanical Engineering, 238*(3), 1060–1071.
75. Verma, P. K., Vasudev, H., Singh, G., & Singh, S. (2024). A review on the microstructural and biomedical properties of implants manufactured using additive manufacturing. *Journal of The Institution of Engineers (India): Series D, 106*, 1–15.
76. Singh, S. P., Verma, P. K., Singh, S., Kapoor, M., & Balu, R. (2024). Microstructural characterization and electrochemical behaviour of HVOF sprayed WC–Co–Cr coatings on 316L boiler steel stainless steel. *Results in Surfaces and Interfaces, 16*, 100248.

6 Applications of Tribological Principles in Additive Manufacturing Technologies

Amrinder Mehta, Hitesh Vasudev, Jashanpreet Singh, and Brijesh Prasad

6.1 INTRODUCTION

The technology of additive manufacturing (AM), or 3D printing, has significantly influenced the production of multifaceted components in different sectors in that complex geometries that were formerly inaccessible or expensive to manufacture with conventional technology are now accessible due to 3D printing. The technology enables the manufacturing of complicated structures with unmatched personalization and decreased waste, improving the output of the production procedure [1]. Advanced designs and geometry constraints have been broken with AM technologies such as digital light processing (DLP), and this has transformed the sector. The possibilities of AM technologies are also spread to the biomedical sphere, where they can be utilized to produce patient-specific structures promoting tissue development and providing an opportunity to customize to a certain extent [2].

This development is an indication of the potential of AM in precision medicine and population health, where designing unique elements is priceless. Furthermore, the ability of AM to produce decent functional electronic elements has enabled embedded circuits to be created using passive elements such as resistors and capacitors directly in the printed form. This integration does not only help in assisting the miniaturization of electronic devices but also increases the functionality of composite components in more applications [3]. The layer-by-layer format of AM allows a freedom of design in a way never seen before, enabling the use of a wide range of materials. Even with the enormous opportunities provided by AM, there are still issues, especially on the aspect of computational design methods that are crucial towards coming up with objects that have customized attributes and functionality [4]. Additive manufacturing of metals has shown considerable promise in particle accelerator use, with capabilities matching those of traditionally produced

DOI: 10.1201/9781003741152-6

counterparts in the way they combine complex internal structures. However, the incorporation of AM into this area has difficulties with laying down trustworthy qualification measures and the functionality of the produced elements in harsh environments [5]. Machine learning (ML) and artificial intelligence (AI) find their way to the center of tribology, assisting in controlling intricate data patterns and trends in various fields such as composite material, drive technology, manufacturing, and surface engineering. Artificial neural networks, decision trees, and support vector machines, among others, are some of the algorithms employed to improve research and development in these areas [6].

Figure 6.1 illustrates a review of the fields of tribology research in two pie charts. The bigger pie chart divides tribology into four broad areas, where intelligent tribology constitutes half of the study (50%), basic theory of tribology constitutes 27%, basic component tribology constitutes 9%, and other fields cover 14%. The inset pie chart also gives a more detailed split of the section of basic component tribology by revealing that under this section, biotribology is the biggest (36%), then extreme tribology (29%), other domains (29%), and finally the smaller share of green tribology (7%). The illustration demonstrates that there is a great emphasis on intelligent and theoretical subjects in the study of tribology, as such specialized studies as biotribology and extreme tribology are being given more priority in component studies. The specific listings and values contribute to the explanation of the relative focus on the different subfields, offering insight into the current research priorities in the field of tribology [7].

Nevertheless, tribology performance, i.e., friction, wear and lubrication, of AM components is a major challenge to their use in the arduous industrial operations. It is a combination of surface finish, choice of material, and the layer-by-layer addition of material that constitutes AM that gives a significant effect on the tribological behavior of printed parts. Recent studies have indicated better process parameter optimization and use of superior materials [8]. Underfitting is the case when a model is too simple to represent underlying patterns of data. This is usually because of the

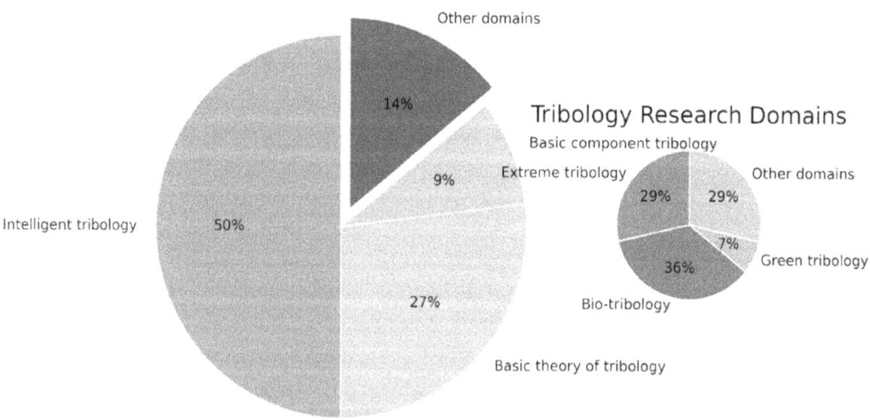

FIGURE 6.1 Proportion of AI technology in various research fields of tribology [7].

utilization of models that are not complex enough, like the use of linear models on nonlinear data. Poor performance of models that are underfit normally applies to training and test data, resulting in high prediction errors. It means that the model has not made sufficient use of the data to learn [9]. Maximum fitting is made possible when a model has a balance of complexity and generalization capacity, and it is effective in representing the underlying structure of the data without overfitting. This refers to a model that is capable of accurate prediction of both training and unseen data. The ideal level of complexity of the model at this stage can be obtained, which generates few errors during training and test data. It is commonly done through the application of validation methods and complexifying the model to prevent both bias and variance [10]. Overfitting occurs when a model picks up the noise and other irrelevant information in training data, and thus the model achieves a high score on training data but not on new unseen data. Overfitting models are associated with low training errors and high test errors. They do not work on new data because they are too complex and based on certain trends in the training data that are not generalizable [11].

Figure 6.2 comprehensively explains underfitting, optimum fitting, and overfitting in machine learning using both regression and classification examples. For regression, underfitting is shown as a straight line failing to capture data trends (high bias), resulting in high training and test errors. Optimum fitting balances complexity, leading to low errors in both sets. Overfitting is depicted as a convoluted curve fitting every data point, yielding low training error but high test error due to high variance. For classification, underfitting misclassifies many points due to an overly simple boundary, while optimum fitting accurately separates classes. Overfitting introduces an overly complex boundary, tailoring too closely to training data noise. Solutions suggested include increasing model complexity or features for underfitting and applying regularization or acquiring more data for overfitting. The bottom plot shows that training error decreases steadily with iterations, but test error first falls then rises after reaching an optimal point, highlighting the sweet spot between underfitting and overfitting for best model generalization [12].

This review critically examines the current state of knowledge on how tribological principles are applied in AM, focusing on the impact of surface finish, material selection, and construction methods, as well as strategies for enhancing the tribological properties of 3D-printed parts. Tribological principles are also important in the prediction and optimization of coating characteristics in the framework of thermal spray processes. These principles are used to assess and enhance the wear resistance and friction qualities of friction of coatings [13]. The wear resistance and hardness of thermal spray coatings depend heavily on the type of material and microstructural features of the material. As an example, web coatings, such as WC–12% Co and Cr_3C_2–MCr, have varied tribological properties when deposited by other methods of thermal spray such as the plasma spray or detonation gun spray. In most cases, detonation gun sprayed coatings are thicker and denser, which gives them better wear resistance as compared to plasma-sprayed coatings [14]. Preparation of the surface and the right choice of coating deposition are essential elements toward desired tribological properties. Such methods as plasma spraying and detonation gun spraying

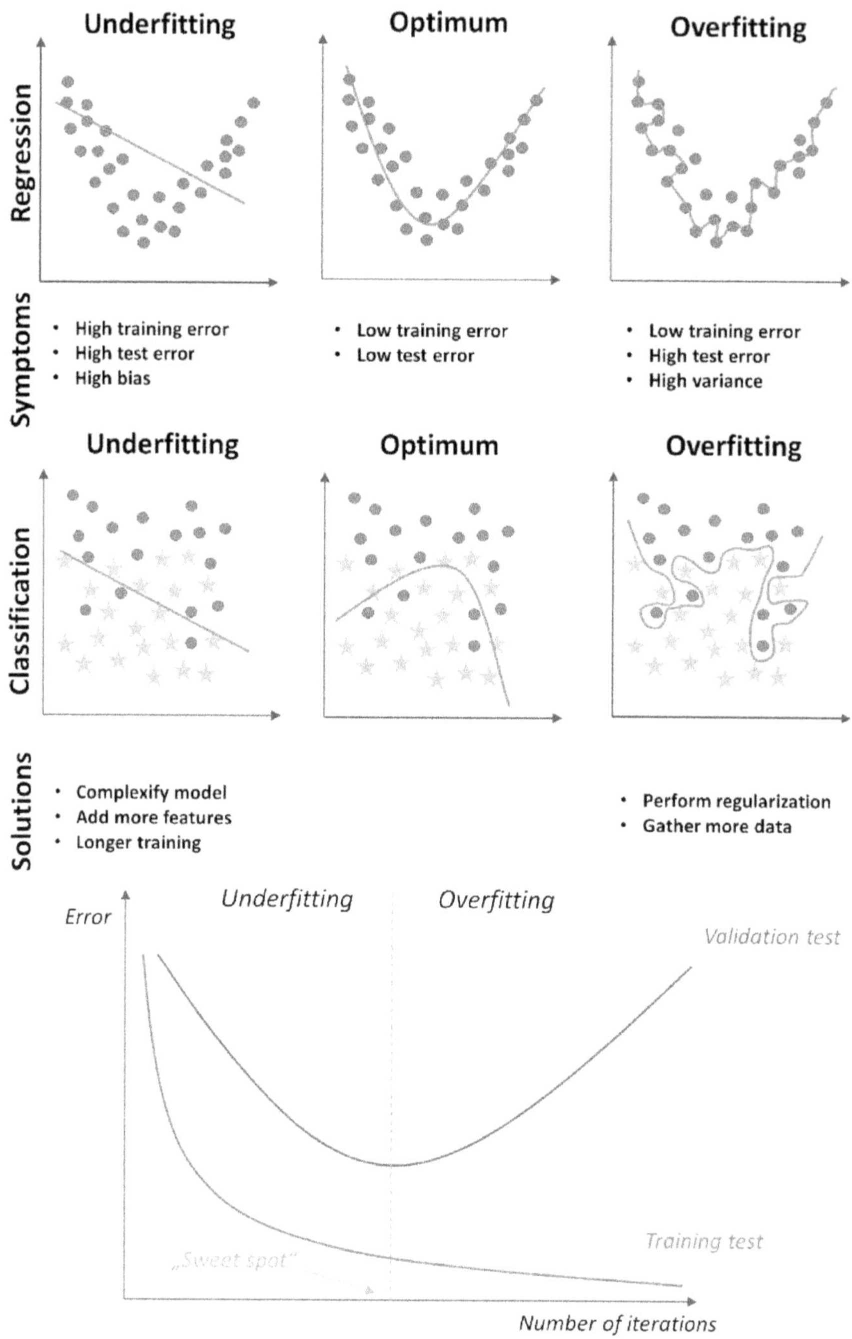

FIGURE 6.2 Examples of underfitting and overfitting [12].

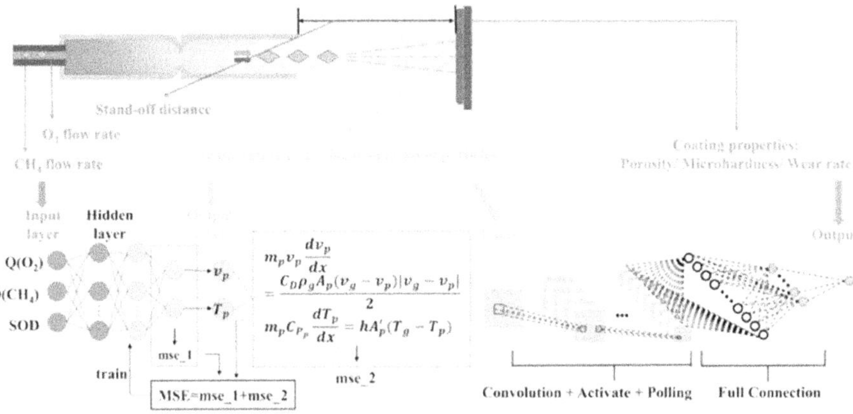

FIGURE 6.3 The prediction of the in-flight particle properties and mechanical performance [16].

influence the porosity of coatings and adhesive properties and their functioning under wear conditions [15].

Figure 6.3 depicts a composite modeling scheme in coating property prediction in a thermal spray procedure. The rate of oxygen and methane inputs, and the distance between the stand-off are introduced into a primary neural net to arrive at the temperature and velocity of particles in transit. These estimated values are further subjected to a series of partial differential equations that model the physics of particle dynamics and heat exchange with the calculated errors summed up by the mean squared error. The resultant physical parameters are the input in a convolutional neural network that undergoes the convolution, activity, and pooling steps, and then the fully connected layers. Finally, the model results in the generation of important coating properties such as porosity, microhardness, and wear rate. It is shown in the workflow that experimental conditions and the laws of physics are incorporated with deep learning architectures to improve the precision of the prediction and the ability to learn intricate relationships between the process parameters and the results of coating in thermal spray technology [16].

There are different tribological tests applied to test coatings, including pin-on-disk sliding wear tests, solid particle erosion tests, and rubber wheel sand abrasion tests. They are used to study how coatings can perform in various stress modes and stress conditions, and this knowledge gives information on how the coating can be used and in what situations it can last [17]. Current developments in the field of coating design have been on the incorporation of hard ceramics and PVD in extreme temperature applications to improve the tribological behavior of the coating. This involves the optimization of such parameters as input power, gas flow rates, and substrate bias to produce coatings that have better hardness, wear resistance, and stability at high temperature [18].

Tribology principles also include the knowledge of lubrication mechanisms, particularly in coatings doped with materials such as PTFE or graphene that have a huge effect on decreasing friction and wear at high temperatures. The applications of liquid lubrication are impossible, and this type of enhancement is essential. The tribological principles applied in the thermal spray processes include an overall knowledge of the material of choice, methods of coating, surface preparation, and test methods to forecast and control the behavior of the coating [19]. Such principles ensure that the coatings can endure different operational challenges in addition to being able to offer better wear resistance and frictional properties.

6.2 IMPACT OF SURFACE FINISH

Surface finish plays an important role in determining the friction and wear properties of the components produced through AM. The coarseness of the layer-by-layer deposition may lead to larger values of friction coefficients and wear than actual components that are manufactured conventionally. As an example, AM parts made of polymer, with reinforcement in terms of copper and aluminum, have been demonstrated to become rougher and more subject to higher coefficients of friction, despite the ability of this reinforcement to increase hardness and decrease wear under some circumstances [20]. Post-processing of AM parts is important in the enhancement of surface quality and wear resistance. As an illustration, the finish machining and drag finishing operations could reduce the surface roughness of Inconel 718 specimens prepared using selective laser melting to a considerable degree. This loss of roughness is associated with a significant drop in the coefficient of friction as well as the wear rate, thereby improving the overall wear performance of the material [21].

Moreover, the tribological performance of polyamide surfaces is dependent on their texture and mode of production. CNC micromachining is also likely to result in a superior surface finish and hardness than 3D printing can, influencing the coefficient of friction and wear resistance [22]. On a similar note, surface treatment of AM parts, e.g., coating with tribological filaments, may also affect surface stability and moderate the coefficient of friction, even though large effects in the coefficient may not be observed at varying coating thicknesses, underscoring the complicated nature of the interaction between surface texture and wear behavior [23].

Figure 6.4 shows an inverse design process of metasurfaces by incorporating a model of deep neural networks (DNNs) and genetic algorithms (GAs). This requirement is the input to the DNN–GA model, where the genetic algorithm is repeatedly used to generate and evolve populations of candidate meta-atom structures, with variation and crossover operations performed on each design to enhance the fitness. The DNN accelerates the optimization through the fast assessment of a generated candidate. On the right of the figure, the meta-atom configurations of interest are written out, each of them adjusted to give the desired phase and amplitude response. The end result of the inverse design process is a metasurface that can be used to control the incident electromagnetic waves into the desired functionality, e.g., producing desired beams shapes [24].

Moreover, it has also been demonstrated that other processes like sliding friction burnishing can decrease the surface roughness and enhance the surface hardness of selective laser-melted Ti6Al4V parts to enhance their tribological behavior and wear

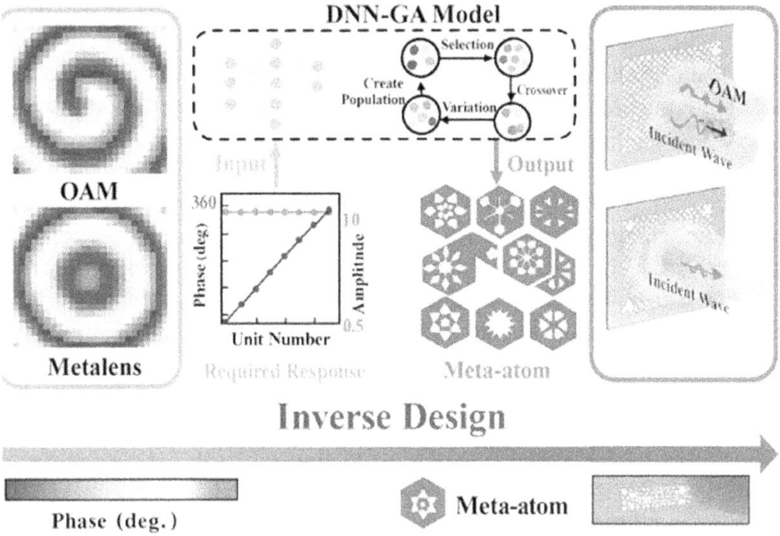

FIGURE 6.4 A schematic of the proposed inverse design [24].

resistance. The processing of Al-Si12 components using friction stir technology was proved to alter microstructural and mechanical properties, improving the surface properties and corrosion resistance [25]. Although the roughness of AM parts is inherent and may result in a higher friction and wear rate, post-processing methods and surface treatments may be used to enhance the tribological performance of this material. The innovations play a vital role in increasing the use of AM technologies in the industries where wear resistance is of significant interest. Surface roughness can be minimized by optimizing printing parameters including layer thickness. The wear resistance and friction are improved further by post-processing abilities such as surface texturing, coatings (e.g., diamond-like carbon), and thermal processes [26].

6.3 MATERIAL SELECTION AND COMPOSITES

The material selection is critical in obtaining certain tribological characteristics where AM processes are used. Various materials such as metals, polymers, ceramics, and composites show different friction and wear results that can be attributed to their properties and the AM processes used [27].

Within the context of metal additive manufacturing, such processes as friction stir additive manufacturing (FSAM) are observed to be characterized by better mechanical properties because of their solid-state character. The process results in components that have equiaxed grains and hence have enhanced mechanical properties and reduced defects as opposed to traditional melting-based processes. Additive friction stir deposition techniques have the benefit of a fully dense material with fine microstructures, which are suitable for high loads. Also, anti-wear effects can be optimized by incorporating electric currents in the manufacturing process like in the case of silver coatings because dislocation and grain refinement processes can be optimized [28]. AM parts made of polymer exhibit high improvements in material reinforcements. As an example, a PLA matrix containing metallic particles such as copper and aluminum makes it harder and wears less, although it causes the surface roughness and coefficient of friction to go up in some instances. Therefore, reinforced polymers will be useful in applications where high stability and durability are needed [29]. AM facilitates eliminating the natural drawback of processing ceramics to be used in tribology. The accuracy and fabrication of AM are especially beneficial in the production of complicated geometries and incorporation of structural and functional elements, which are difficult to achieve through traditional processes. Ceramics have better wear resistance and temperature stability in the workplace and can be used in industries that require high performance [30]. The creation of metal matrix composites (e.g., with aluminum), the creation of which may use processes like MELD, demonstrates potential to create high-strength parts without the issue of porosity. The kind of reinforcement particles and matrix material play an important role in determining the wear behavior and mechanical characteristics and, therefore, it is essential to customize composite compositions to fit the application [31]. Tribological properties are highly specific to AM material selection and processing; thus, it is an important field of current research and technology development. Although AM technology allows obtaining complex and high-performance components, issues including the optimization of material compositions and processing parameters remain one of the subjects of ongoing study [32].

The anisotropic characteristics of AM greatly arise from the layer-by-layer construction and influence the tribological characteristics of the parts made. These characteristics may change depending on the direction of wear because of the specific anisotropic properties of the materials, which are a consequence of the used manufacturing methods [33].

An investigation of selective laser melting (SLM) created AlSi10Mg alloys; wear anisotropy illustrated how varying scanning strategies and orientations affect the wear and friction characteristics manifold. Depending on the configuration of the scanning vectors and design blueprints used in the fabrication, samples exhibited disparities in wear and friction coefficients. Additionally, the samples' friction wear rates and coefficients were affected by whether fresh or recycled powder was utilized [34]. The material anisotropy presented by additively manufactured polymers and composites highlighted how much the mechanical characteristics—specifically, the tensile and shear strengths and tribological properties—can vary with the building direction and raster orientation. This anisotropy makes it difficult to achieve consistency in the mechanical and tribological properties of the different regions of a part [35].

Moreover, the tribological performance of additively manufactured parts is influenced by the surface condition. For instance, surface treatments such as grinding and polishing add value to tribological performance by endowing additively manufactured AISI H13 steel with a lower friction coefficient than that of the as-printed condition. Such surface conditions are important since the friction and wear properties are determined by the interaction of the friction with the layer-by-layer anisotropic structure [36]. AM's anisotropic nature leads to considerable design complexities associated with each of the manufacturing and post-processing treatments. This leads to difficulties in the optimization of the tribological properties of additively manufactured parts. This complexity in optimization should be a focus of research to aid in the full utilization of AM technologies and to reduce the associated complications due to anisotropic properties [37].

Figure 6.5 shows the direction in which the black-box deep learning solution, which is typically employed in the analysis of medical images, can be converted into a clinically acceptable and interpretable model. It starts with input images that are interpreted by an opaque deep neural network, the decision-making of which is not clear. To increase interpretability, different approaches are proposed: concept learning models, case-based models, counterfactual explanations, concept attribution, language description, latent space interpretation, attribution maps, anatomical priors, and analysis of internal network representations. Such methods are useful to explain predictions and associate them with clinical concepts. The interpretability techniques are then measured by clinical environments, imaging biomarkers, adversarial robustness, and quantitative measures of trustworthy explanations. Once this multistep validation process is complete, the deep learning solution becomes transparent and reliable for clinical use, which is represented as an open-box model with transparent reasoning. The flow highlights the need to explain and trust in the use of AI-powered diagnostics or decision support in healthcare [38].

There are also a variety of factors that make vertically oriented parts have a high friction and a low wear depth relative to their horizontally oriented counterparts. The contact area can have dissimilar pressure distributions in vertical orientations as compared to horizontal orientations. Friction may increase with higher normal loads that are caused by gravitational forces. These loads could also, however, cause more even distributions of wear, decreasing the depth of localized wear. The grain structure may be in such a way that it increases the resistance to material removal

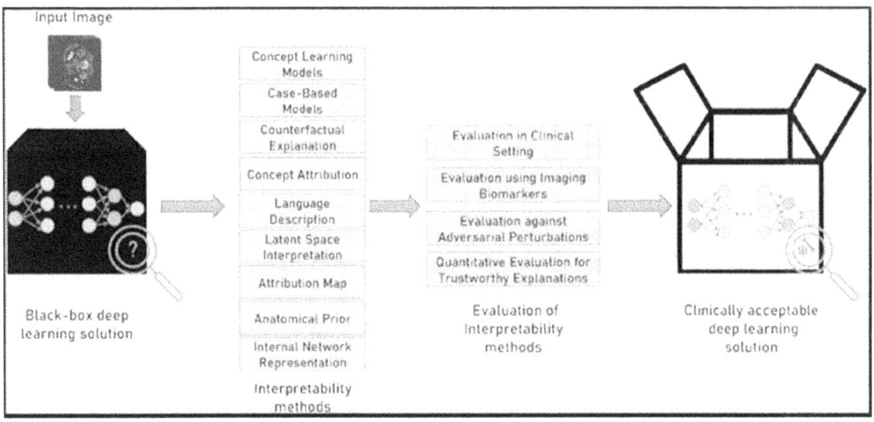

FIGURE 6.5 Overview of interpretability methods for deep learning solutions in medical image analysis [38].

leading to the reduction of the wear depths [39]. The orientation will also influence the lubrication conditions; the directional lubricants may stay on the vertically oriented surface owing to the effect of gravity resulting in less wear. This, though, may not play a crucial role in the minimization of friction if the roughness of the surface and interactions of materials still take the foreground. The effect of orientation on the heat dissipated during frictional contact can take place. Vertical orientations may be more advantageous allowing heat to be dissipated more effectively, keeping the contact surface cooler and possibly preventing material degradation due to thermal effects as well as preventing normal wear due to mechanical loading [40]. The parts that have a vertical orientation may respond differently to the environment like dust and moisture than a horizontally oriented part. The interactions may have varying effects on wear and friction properties according to orientation depending on the variation in contaminant deposition and retention of moisture. A combination of these factors tends to lead to the seen behavior that has vertically oriented parts more deeply frictional and less deep wear when compared to horizontally oriented parts [41].

Figure 6.6 gives a general description of the key branches of machine learning, namely, supervised, unsupervised, and reinforcement learning. In supervised learning, the models are trained on annotated data; one can mention support vector

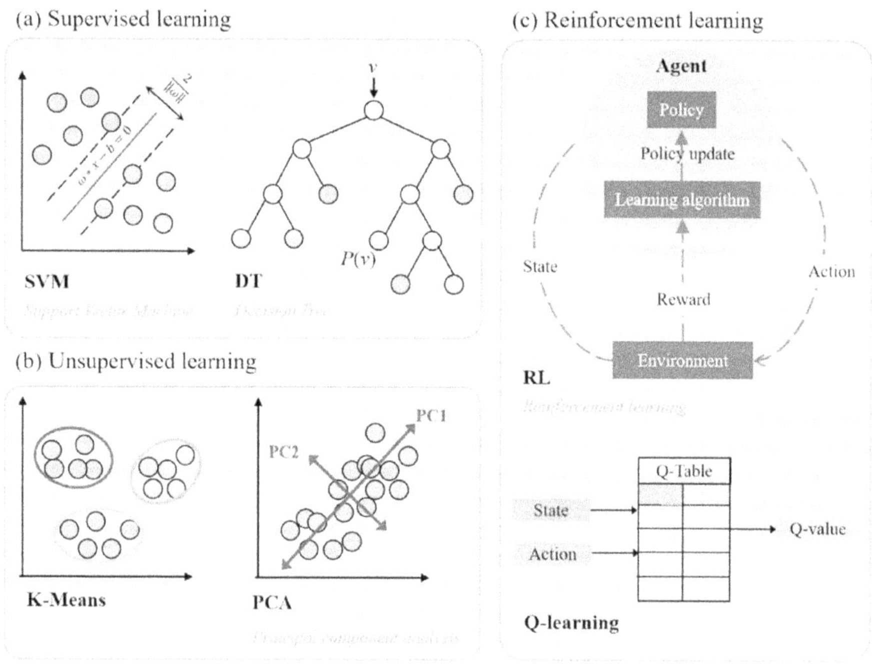

FIGURE 6.6 (a) Supervised learning models SVM and DT, (b) unsupervised learning models K-means and PCA, and (c) framework of reinforcement learning and Q learning mode [42].

machine (SVM), which finds an optimal boundary between the categories, and decision tree (DT), which makes stepwise splits in the data, depending on the values of features. Unsupervised learning is concerned with data presented in an unlabeled form; its aim is to recognize patterns and dimensionality reduction. An example of this is the K-means clustering (presenting data in clusters) and principal component analysis (PCA) (presenting data along major axes of feature extraction). The characteristics of reinforcement learning include the agent that interacts with the environment, makes decisions (actions), and receives feedback (rewards) to enhance its policy in the long run. The structure consists of a Q-table where learned values of state–action pairs are stored, which results in optimal future actions. These diagrams allow summarizing the most important paradigms and algorithms of machine learning applied in data-driven research and intelligent systems [42].

When optimally done, AM has already proven capable of producing and even surpassing the tribological characteristics of traditionally manufactured parts by far. Different research points to the improvements in the tribological behavior of AM components through material improvements, processing methods, and surface texturing. In the case of polymer AM, it has been observed that the incorporation of reinforcement materials, which include copper and aluminum in a PLA matrix, greatly leads to better tribological performance in terms of hardness and wear rates. Likewise, Inconel 625 superalloys synthesized through AM of electron beam selective melting had better mechanical properties and excellent wear resistance than the conventional process because of their strength enhancement and the protection tribo-layers [43].

Figure 6.7 compares the traditional neural network method and the physics-based method of neural network in predicting film thickness in a system. The left panel describes the physics-informed neural network (PINN) workflow, which begins with the process of initializing, adjusting the parameters, and repeating the process of finding solutions until the necessary load balance of the solution is achieved. The main figure reports the experiment of film thickness versus velocity revealing that

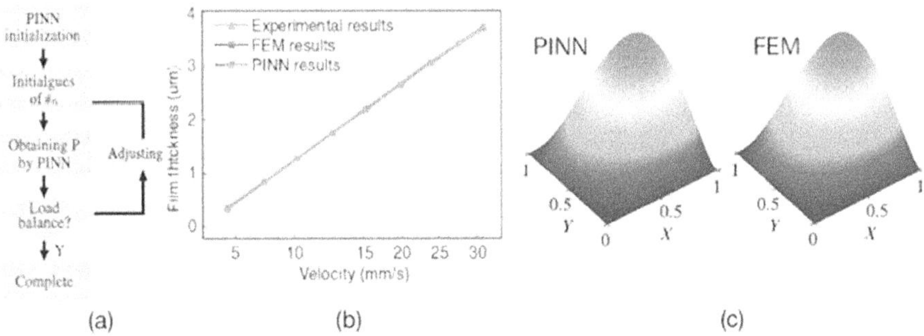

FIGURE 6.7 (a) Flowchart of iterative PINN approach for hydrodynamic contact. (b) Outlet film thickness at different sliding velocities for the PINN method compared with FEM simulation and experimental results. (c) Pressure distribution predicted using PINN (left) and FEM (right) [44].

PINN findings, the finite element method (FEM), and experimental findings are similar, proving the reliability of the PINN methodology. The right side shows two 3D surface plots, one with PINN and another one with FEM, both with the distribution of the film thickness in the X-Y plane. The color gradients and shapes of the surfaces are also almost the same, supporting the concurrence of the deep learning and the traditional simulation methods. The big picture is that more modern neural networks are capable of matching the accuracy of traditional computations and implementing solutions in a short period of time and through an iterative process influenced by physical laws [44].

Figure 6.8 shows that neural network augmented by a data-driven optimizer can be used to simulate the dynamics of friction and wear contacts. At the head, there is a schematic illustrating two solids coming into contact, which characterizes regions of pressure inside the domain of lubricants and the solid–solid interface. Input data including time are entered under four multilayer perceptron models specifically focused on wear rate, solid pressure, hydrodynamic pressure, and friction with several hidden layers and neurons each. The integration of their results is based on a loss in which variations in predicted and actual variables are considered. Through this combined method, a number of tribological factors can be predicted and optimized simultaneously to provide detailed analysis and enhanced accuracy of complex contact phenomena in machine learning-based simulations of friction and wear [45].

Green tribology principles are in line with AM in ensuring sustainable manufacturing processes. The AM technologies can also help to promote the environmentally friendly and efficient production process by reducing friction and wear, and developing self-lubricating materials and natural lubricants [46].

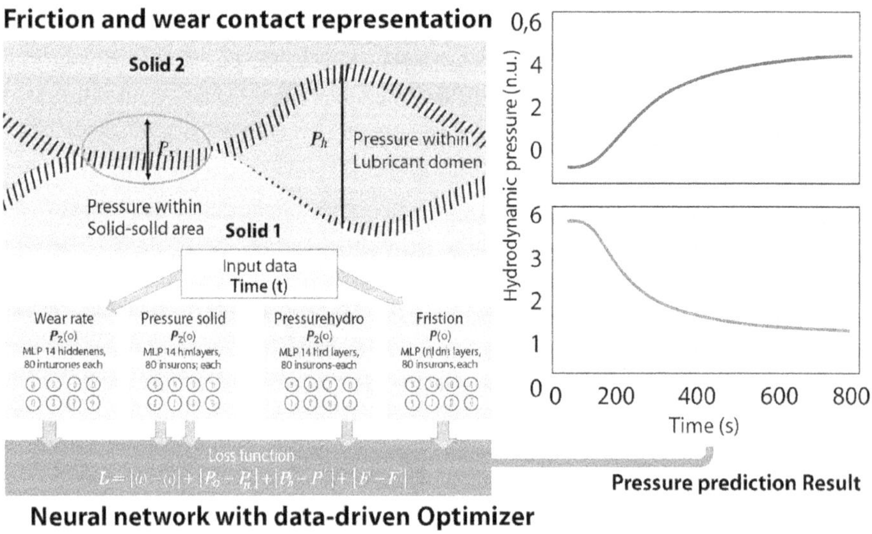

FIGURE 6.8 Plate wear condition prediction based on ANN with data-driven optimization [45].

Engineers from all walks of life can use coating technologies to look at surfaces and structures in new ways. They include tests for surface roughness, tests for mechanical integrity (including tensile, compressive, and flexural strengths), and tests for ways to improve solar energy systems' optical and thermal qualities, which lead to better performance. Coatings also help to depict fluid movement in difficult-to-understand networks of channels and transitions, such as elbows and pipelines. Coatings with specific functions make it easier to foresee how stresses would affect a structure and how it will function in a precision-engineered system, such as an ergonomically sensitive seating arrangement. Because of this, the systems last longer, the designs are better, and multiphysics simulations are less prone to mistakes.

6.4 CONCLUSION

This review examines the interface between tribological studies and the emerging technologies in AM. The friction and wear characteristics of AM components are influenced by the surface finish, the material composition, and the layer-by-layer assembly approach. Compared to conventional parts, AM components with rough surfaces exhibit higher friction and wear. This scenario may improve after surface finishing techniques like machining, polishing, and surface treatments. Friction and wear characteristics differ because of the varying component materials used (metals, polymers, and composites). The wear resistance of polymer matrices is significantly enhanced by incorporating wear-resistant composites and fine metallic particles, thereby also improving wear resistance. The anisotropic nature of parts produced by the layer-by-layer construction method results in friction and wear properties being altered. The build orientation and raster patterns are determinants of friction and wear due to anisotropic construction patterns. The future prediction and optimization of tribological performance of AM components is enhanced with the integration of machine learning and physics-informed neural networks. These developments are aided by the AM techniques and principles of sustainable tribology. There will be improvements in the tribological performance of AM components, especially in the applications required by the industry, with further process optimization, novel materials, and enhanced post-processing methods.

REFERENCES

1. O. Abdulhameed, A. Al-Ahmari, W. Ameen, S. Mian, Additive manufacturing: Challenges, trends, and applications, *Advances in Mechanical Engineering* 11 (2019) 1–27.
2. A.M. Aboutaleb, L. Bian, A. Elwany, N. Shamsaei, S.M. Thompson, G. Tapia, Accelerated process optimization for laser-based additive manufacturing by leveraging similar prior studies, *IISE Transactions* 49(1) (2017) 31–44.
3. L. Adam, O. Lietaer, S. Mathieu, I. Doghri, Numerical simulation of additive manufacturing of polymers and polymer-based composites, in *Structure and Properties of Additive Manufactured Polymer Components*, Elsevier, 2020, pp. 115–146.
4. S.K. Adapa, Jagadish, Prospects of natural fiber-reinforced polymer composites for additive manufacturing applications: A review, *JOM* 75 (2023) 920–940.

5. M. Afrasiabi, M. Bambach, Modelling and simulation of metal additive manufacturing processes with particle methods: A review, *Virtual Physical Prototyping* 18(1) (2023) e2274494.
6. L. Agnusdei, A. Del Prete, Additive manufacturing for sustainability: A systematic literature review, *Sustainable Futures* 4(5) (2022) 100098.
7. N. Yin, P. Yang, S. Liu, S. Pan, Z. Zhang, AI for tribology: Present and future, *Friction* 12(6) (2024) 1060–1097.
8. D.-G. Ahn, Direct metal additive manufacturing processes and their sustainable applications for green technology: A review, *International Journal of Precision Engineering Manufacturing-Green Technology* 3 (2016) 381–395.
9. H. Lindroth, K. Nalaie, R. Raghu, I.N. Ayala, C. Busch, A. Bhattacharyya, P. Moreno Franco, D.A. Diedrich, B.W. Pickering, V. Herasevich, Applied artificial intelligence in healthcare: A review of computer vision technology application in hospital settings, *Journal of Imaging* 10(4) (2024) 81.
10. F. Anifowose, A. Khoukhi, A. Abdulraheem, Investigating the effect of training–testing data stratification on the performance of soft computing techniques: an experimental study, *Journal of Experimental & Theoretical Artificial Intelligence* 29(3) (2017) 517–535.
11. J.M. Kernbach, V.E. Staartjes, Foundations of machine learning-based clinical prediction modeling: Part II—generalization and overfitting, in: V.E. Staartjes, L. Regli, C. Serra (Eds.) *Machine Learning in Clinical Neuroscience*, Springer International Publishing, Cham, 2022, pp. 15–21.
12. Ł. Ledziński, G. Grześk, Artificial intelligence technologies in cardiology, *Journal of Cardiovascular Development and Disease* 10(5) (2023) 202.
13. F. Alifui-Segbaya, I.F. Ituarte, S. Hasanov, A. Gupta, I. Fidan, Opportunities and limitations of additive manufacturing, in: *Springer Handbook of Additive Manufacturing*, Springer, 2023, pp. 125–143.
14. A.B. Elshalakany, V. Tirth, M. Elsisy, B. Alqahtani, R.A. El-Sehiemy, V. Amigó, W. Hozief, A. Algahtani, W. El-Sallamy, performance study of MCrAlY flame sprayed coating on AISI 304 stainless steel of microstructure and sliding wear at high temperatures, *JOM* 77(2) (2025) 549–563.
15. Anahita Shamsini, Mohsen Saremi, Mahmoud Heydarzadeh Sohi, Oxidation and hot corrosion characteristics of YSZ/ NiCoCrAlY electrophoretically deposited on IN-738, *Journal of Materials Research and Technology*, 41 (2026) 1193–1203.
16. L. Gui, B. Wang, R. Cai, Z. Yu, M. Liu, Q. Zhu, Y. Xie, S. Liu, A. Killinger, Prediction of in-flight particle properties and mechanical performances of HVOF-sprayed NiCr–Cr3C2 coatings based on a hierarchical neural network, *Materials* 16(18) (2023) 6279.
17. S. Kaushal, D. Gupta, H. Bhowmick, Investigation of dry sliding wear behavior of Ni–SiC microwave cladding, *Journal of Tribology* 139(4) (2017) 041603.
18. J.A. Barreras-Uruchurtu, N. Besnard, C. Paul, L. Marchal, S. Devisme, B. Lestriez, Effect of PVdF distribution on properties and performance of dry spray-coated graphite electrodes for lithium-ion batteries for electric vehicle applications, *Journal of The Electrochemical Society* 171(8) (2024) 080511.
19. F. Wang, H. Zhu, H. Zhang, H. Tang, J. Chen, Y. Guo, Effect of surface hydrophilic modification on the wettability, surface charge property and separation performance of PTFE membrane, *Journal of Water Process Engineering* 8 (2015) 11–18.
20. S.M. Ansari P, On the selective laser melting based additive manufacturing of AlSi10Mg: The process parameter investigation through multiphysics simulation and experimental validation, *Journal of Alloys Compounds* 890 (2022) 161873.

21. M. Alizadeh-Sh, S. Marashi, E. Ranjbarnodeh, R. Shoja-Razavi, Laser cladding of Inconel 718 powder on a non-weldable substrate: Clad bead geometry-solidification cracking relationship, *Journal of Manufacturing Processes* 56 (2020) 54–62.
22. R.A. Chowdhury, S.X. Peng, J. Youngblood, Improved order parameter (alignment) determination in cellulose nanocrystal (CNC) films by a simple optical birefringence method, *Cellulose* 24 (2017) 1957–1970.
23. I.F. Edem, P.T. Mativenga, Modelling of energy demand from computer numerical control (CNC) toolpaths, *Journal of Cleaner Production* 157 (2017) 310–321.
24. G. Wu, L. Si, H. Xu, R. Niu, Y. Zhuang, H. Sun, J. Ding, Phase-to-pattern inverse design for a fast realization of a functional metasurface by combining a deep neural network and a genetic algorithm, *Optics Express* 30(25) (2022) 45612–45623.
25. L. Bai, J. Li, J. Chen, R. Song, J. Shao, C. Qu, Effect of the content of B4C on microstructural evolution and wear behaviors of the laser-clad coatings fabricated on Ti6Al4V, *Optics Laser Technology* 76 (2016) 33–45.
26. M. Bai, L. Reddy, T. Hussain, Experimental and thermodynamic investigations on the chlorine-induced corrosion of HVOF thermal sprayed NiAl coatings and 304 stainless steels at 700 C, *Corrosion Science* 135 (2018) 147–157.
27. K. Derelizade, F. Venturi, R. Wellman, A. Kholobysov, T. Hussain, Wear performance of graphene nano platelets incorporated WC-Co coatings deposited by hybrid high velocity oxy fuel thermal spray, *Wear* 482 (2021) 203974.
28. H.A. Derazkola, F. Khodabakhshi, A. Simchi, Evaluation of a polymer-steel laminated sheet composite structure produced by friction stir additive manufacturing (FSAM) technology, *Polymer Testing* 90 (2020) 106690.
29. P. Parandoush, D. Lin, A review on additive manufacturing of polymer-fiber composites, *Composite Structures* 182 (2017) 36–53.
30. M.S. Hasan, A. Kordijazi, P.K. Rohatgi, M. Nosonovsky, Triboinformatic modeling of dry friction and wear of aluminum base alloys using machine learning algorithms, *Tribology International* 161 (2021) 107065.
31. V. Hassija, V. Chamola, A. Mahapatra, A. Singal, D. Goel, K. Huang, S. Scardapane, I. Spinelli, M. Mahmud, A. Hussain, Interpreting black-box models: A review on explainable artificial intelligence, *Cognitive Computation* 16(1) (2024) 45–74.
32. P. Avrampos, G.-C. Vosniakos, A review of powder deposition in additive manufacturing by powder bed fusion, *Journal of Manufacturing Processes* 74 (2022) 332–352.
33. P. Awasthi, S.S. Banerjee, Fused deposition modeling of thermoplastic elastomeric materials: Challenges and opportunities, *Additive Manufacturing* 46 (2021) 102177.
34. Y. Bai, C. Zhao, J. Yang, J.Y.H. Fuh, W.F. Lu, C. Weng, H. Wang, Dry mechanical-electrochemical polishing of selective laser melted 316L stainless steel, *Materials & Design* 193 (2020) 108840.
35. M. Benedetti, E. Torresani, M. Leoni, V. Fontanari, M. Bandini, C. Pederzolli, C. Potrich, The effect of post-sintering treatments on the fatigue and biological behavior of Ti-6Al-4V ELI parts made by selective laser melting, *Journal of the Mechanical Behavior of Biomedical Materials* 71 (2017) 295–306.
36. M. Dai, Y. Jiang, F. Yang, X. Xu, W. Zhao, M.H. Dao, Y. Liu, SLMGAN: Single-layer metasurface design with symmetrical free-form patterns using generative adversarial networks, *Applied Soft Computing* 130 (2022) 109646.
37. C. Han, Y. Li, Q. Wang, D. Cai, Q. Wei, L. Yang, S. Wen, J. Liu, Y. Shi, Titanium/hydroxyapatite (Ti/HA) gradient materials with quasi-continuous ratios fabricated by SLM: material interface and fracture toughness, *Materials Design* 141 (2018) 256–266.
38. Z. Salahuddin, H.C. Woodruff, A. Chatterjee, P. Lambin, Transparency of deep neural networks for medical image analysis: A review of interpretability methods, *Computers in Biology and Medicine* 140 (2022) 105111.

39. G. Han, Z. Wang, K. Liu, S. Li, X. Du, W. Du, Synthesis of CNT-reinforced AZ31 magnesium alloy composites with uniformly distributed CNTs, *Materials Science and Engineering: A* 628 (2015) 350–357.
40. H. Han, C. Li, X. Qiu, Multi-modal remote sensing image matching method based on deep learning technology, *Journal of Physics: Conference Series* 2083 (2021) 032093.
41. H. Han, H. Yu, H. Wei, J. Gong, W. Xu, Recent progress in three-terminal artificial synapses: From device to system, *Small* 15(32) (2019) 1900695.
42. Q. Wang, Y. Li, R. Li, Integrating artificial intelligence in energy transition: A comprehensive review, *Energy Strategy Reviews* 57 (2025) 101600.
43. S. Bhatnagar, S. Mullick, A study on the influence of reinforcement particle size in laser cladding of TiC/Inconel 625 metal matrix composite, *Optics Laser Technology* 161 (2023) 109115.
44. Y. Zhao, L. Guo, P.P.L. Wong, Application of physics-informed neural network in the analysis of hydrodynamic lubrication, *Friction* 11(7) (2023) 1253–1264.
45. A. Kovalev, Y. Tian, Y. Meng, Prediction of ball-on-plate friction and wear by ANN with data-driven optimization, *Friction* 12(6) (2024) 1235–1249.
46. T. Coelli, L. Lauwers, G. Van Huylenbroeck, Environmental efficiency measurement and the materials balance condition, *Journal of Productivity Analysis* 28(1) (2007) 3–12.
47. P.K. Verma, H. Vasudev and V. Kumar, Dental implant materials and designs and failure during implant placement and loading: A review, in: *Advanced manufacturing processes*, CRC Press, 2025, pp.138–150.
48. P.K. Verma, H. Vasudev and V. Kumar, Bioactivity and corrosion performance of low-velocity oxy-fuel-sprayed hydroxyapatite-based coatings on a surgical-grade stainless steel, *Transactions of the Indian Institute of Metals* 78(8) (2025) 185.
49. P.K. Verma, H. Vasudev and V. Kumar, Effect of electro-corrosion behavior of different grades of stainless steel used in bio-implants: A review, in: *Advanced manufacturing processes*, CRC Press, 2025, pp. 269–283.
50. VP.K. Verma and H. Vasudev, Importance of hydroxyapatite coatings in improving new bone osteogenesis: A review, in: *Advanced manufacturing processes*, CRC Press, 2025, pp. 284–299.
51. V. Dutta, L. Thakur, B. Singh and H. Vasudev, A study of erosion–corrosion behaviour of friction stir-processed chromium-reinforced NiAl bronze composite, *Materials* 15(15) (2022) 5401.
52. M. Singh, H. Vasudev and R. Kumar, Corrosion and tribological behaviour of BN thin films deposited using magnetron sputtering, *International Journal of Surface Engineering and Interdisciplinary Materials Science (IJSEIMS)* 9(2) (2021) 24–39.
53. H. Sharanabasava, M. Raviprakash, C.D. Prasad, M.R. Ramesh, M.V. Phanibhushana, VasudevH. and S. Kumar, Microstructure, mechanical and wear properties of SiC and Mo reinforced NiCr microwave cladding, *Advances in Materials and Processing Technologies* 10 (2023) 3620–3633.
54. M. Singh, H. Vasudev and M. Singh, Surface protection of SS-316L with boron nitride based thin films using radio frequency magnetron sputtering technique, *Journal of Electrochemical Science and Engineering* 12(5) (2022) 851–863.
55. G. Prashar, H. Vasudev, L. Thakur and A. Bansal, Performance of thermally sprayed hydroxyapatite coatings for biomedical implants: A comprehensive review, *Surface Review and Letters* 30(01) (2023) 2241001.
56. H. Vasudev, Wear characteristics of Ni-WC powder deposited by using a microwave route on mild steel: Microwave cladding of Ni-WC, *International Journal of Surface Engineering and Interdisciplinary Materials Science (IJSEIMS)* 8(1) (2020) 44–54.

57. J. Parkash, H.S. Saggu and H. Vasudev, A short review on the performance of high velocity oxy-fuel coatings in boiler steel applications, *Materials Today: Proceedings* 50 (2022) 1442–1446.
58. P. Singh, H. Vasudev and A. Bansal, Effect of post-heat treatment on the microstructural, mechanical, and bioactivity behavior of the microwave-assisted alumina-reinforced hydroxyapatite cladding, *Proceedings of the Institution of Mechanical Engineers, Part E: Journal of Process Mechanical Engineering* 238(3) (2024) 1060–1071.
59. P.K. Verma, H. Vasudev, G. Singh and S. Singh, A review on the microstructural and biomedical properties of implants manufactured using additive manufacturing, *Journal of The Institution of Engineers (India): Series D* 106 (2024) 1–15.
60. S.P. Singh, P.K. Verma, S. Singh, M. Kapoor and R. Balu, Microstructural characterization and electrochemical behaviour of HVOF sprayed WC-Co-Cr coatings on 316L boiler steel stainless steel, *Results in Surfaces and Interfaces* 16 (2024) 100248.
61. VP.K. Verma, A.S. Minhas, P. Singh and KumarS. , Slurry erosion behaviour of thermal sprayed Al2O3 and Cr2O3 coatings for turbine steels, AIP Conference Proceedings. AIP Publishing LLC, Vol. 2986, No. 1, p. 020012.
62. P.K. Verma, V. Kumar and H. Vasudev, Bioactivity and corrosion analysis of thermally sprayed hydroxyapatite based coatings, *Journal of Electrochemical Science and Engineering* 14(4) (2024) 419–439.
63. P.K. Verma, H. Vasudev and V. Kumar, Hybrid alumina-reinforced hydroxyapatite coatings deposited by using low-velocity oxy-fuel: Electrochemical corrosion and bioactivity assessment for biomedical applications, *Surface and Coatings Technology* 519 (2025) 132921.
64. P.K. Verma, H. Vasudev and V. Kumar, Biocompatibility and corrosion performance of HA-based LVOF sprayed and post-annealed coatings on SS 254 surgical-grade bioimplant material, *Ceramics International* 51(10) (2025) 13207–13225.
65. P.K. Verma, H. Vasudev and G. Singh, Improving the corrosion performance of SS 304 stainless steel using HVOF-sprayed WC-Co-Cr coatings in chloride environments, *Journal of Bio- and Tribo-Corrosion* 11(4) (2025) 1–13.

7 Leveraging Data Analytics and System Optimization in Tribological Processes

Umesh Ghorpade, Shubhangi Suryawanshi, Digvijay Bhosale, Pranal Patil, Satyajeet Gaikwad, and Gursimran Kaur

7.1 INTRODUCTION

As engineering technology advances, specialized equipment and industrial products become more complex, and their expectation for stability and operating conditions becomes more demanding. Expectation for stability and dependability has also increased at the same time.

The three main components of a tribological system are typically the lubricating medium, the working environment, and the tribo-pairs (Figure 7.1). Coupling effects, time-dependent behaviors, and system-level interaction are characteristics of such systems. Numerous theoretical models and a variety of data sources have been developed in tribological research as a result of these complexities. In order to overcome these obstacles, scientists frequently use engineering data, simulations, and experimental investigations to develop theoretical and predictive models or to explain tribological phenomena.

Essentially, tribology is built on data and models. Improving the information exchange between them is crucial to increasing the effectiveness of research. Since it makes data generation, collection, processing, and analysis possible, information technology is essential to this process. Traditional techniques like regression, curve fitting, and inductive reasoning are no longer adequate due to advancements in technology. The efficiency and reach of information processing in tribology have been greatly increased by the development of artificial intelligence and machine learning.

Furthermore, this development has encouraged the integration of informatics with other domains, including educational informatics, health informatics, and supply chain informatics. In this context, tribo-informatics turns into a specialized field. Tribo-informatics has greatly enhanced the effectiveness and efficiency of tribology standards, databases, and information technology for the collection, classification, archiving, retrieval, analysis, and dissemination of tribology-related data.

DOI: 10.1201/9781003741152-7

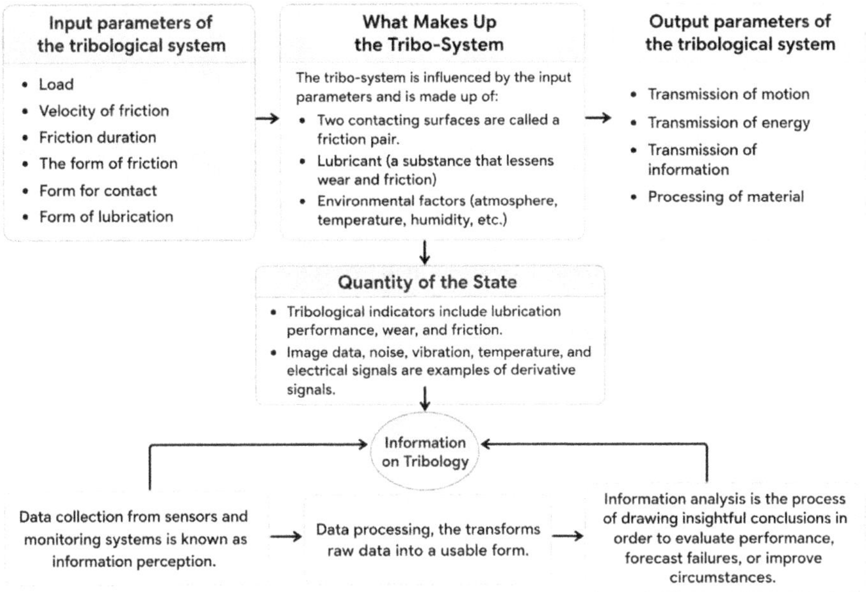

FIGURE 7.1 Components of tribo-system.

These days, tribo-informatics employs a number of methods, such as regression analysis and clustering. Its primary applications include:

1. Condition monitoring—Identifying and assessing the operational state of the tribological system.
2. Behavior prediction—Forecasting a system's performance over time.
3. System optimization—Improving the efficiency and dependability of tribological systems.

The ultimate goal of informatics techniques is to identify relationships in data. Specifically:

- Behavior prediction is used to look for a relationship between tribological parameters and time.
- Condition monitoring is used to determine how tribological parameters and system states relate to one another.
- When analyzing the relationship between systems' inputs and outputs, optimization techniques are used.

One of the main challenges that will enable more precise analysis using informatics techniques is the development of tribological models that accurately attribute different data types. According to this perspective, models direct data processing, while data enhances and validates models. Therefore, essential techniques in tribo-informatics research are the cooperative creation of models and data. The application of informatics techniques in tribology is thus extensively examined in this article, with a focus on the three primary domains of behavior prediction, condition monitoring, and system optimization.

7.2 ROLE OF DATA ANALYTICS IN TRIBOLOGY

7.2.1 Monitoring/Tracking Tribological System Conditions

The process of assessing a tribological system's functional state through the use of indirect or by-product information is known as condition monitoring (Figure 7.2). IT is necessary for both real-time fault diagnosis and mechanical system maintenance.

For example, the operational condition of a bearing can be evaluated using tribological parameters such as friction force and friction torque. This task has been effectively tackled by contemporary machine learning techniques, such as random forest (RF), gradient boosting classifier (GBC), and extra trees classifier (ETC) [1].

Researchers use readily measurable state variables like images, sound pressure, temperature, lubricating oil quality, and vibration signals to indirectly evaluate the tribological condition because it can be difficult to measure tribological variables (such as wear, friction, and lubrication) directly.

7.2.1.1 Monitoring of Wear Status

One of the most obvious and significant elements influencing the lifespan and functionality of a mechanical system is wear. Variation in wear conditions frequently results in audible anomalies in vibration responses, image features, and sound signals [2]. Furthermore, examining the abrasive articles produced during operation offers important information about wear behavior.

Two board categories can be used to specify wear monitoring:

1. Machining part wear monitoring
   Because tool deterioration has a direct impact on machining quality and efficiency, tool wear monitoring is a widely researched application. Accurate

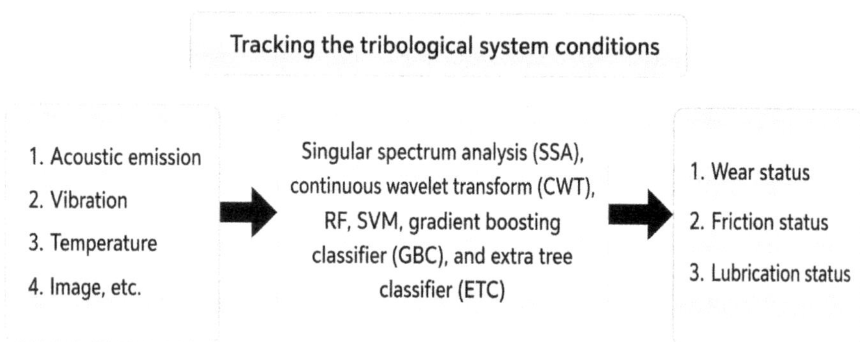

FIGURE 7.2 Tracking tribological system conditions.

scheduling of tool replacement or repair is made possible by real-time monitoring. Images, vibration, and sound pressure are frequently employed signals for tool wear detection.

- Tool wear, phenomena, and acoustic signals can be correlated using the support vector machine (SVM) model.
- Extended regularized particle filtering is one example of a semi-physical model that helps lower error from processing that is solely based on data.
- Wear detection accuracy is increased by advanced machine learning techniques that combine dynamic smoothing with multifeatures and multimodal ensembles.
- Tool wear monitoring has also shown success with pseudo-local singular spectrum analysis (SSA) of vibration signals [3].

The majority of recent studies have focused on the qualitative tools we are monitoring. Thus, adding a decision-making algorithm could improve the accuracy and efficiency of monitoring.

2. Wear tracking of operational components
   Wear is also a common problem in mechanical systems, such as movable connection parts. Keeping an eye on their wear condition yields useful information for evaluating the system's quality and forecasting its remaining service life. Techniques for evaluating degrees of wear based on images have been proposed [4].

- Wear conditions like running-in wear, inadequate lubrication, and particle contamination in oil can be distinguished using the acoustic emission method.
- Wear conditions can be effectively monitored by combining an SVM classifier with continuous wavelet transforms (CWTs).
- Wear states can also be inferred from the morphological characteristics of wear particles. For instance, particle structure and wear behavior have been correlated using the radial concave deviation (RCD) method [5].

7.2.1.2 Monitoring the Friction Status

Monitoring of friction states consists of:

1. Keeping an eye on friction forms
2. Tracking friction characteristic numbers, such as friction, torque, and force

At the moment, there are more techniques for tracking fiction from the friction characteristics [6]. Despite the paucity of research in the field, tracking torque and frictional force is crucial for high precision and high stability systems like satellite attitude control systems (e.g., momentum wheels), and add aero-engine shaft-bush mechanisms.

Friction monitoring will likely become even more crucial as industrial products are expected to meet ever-increasing demands for accuracy and dependability.

7.2.1.3 Monitoring of Lubrication Status

Wear and friction are both generally impacted by lubrication. Two viewpoints are typically used when monitoring the condition of the lubricant:

1. Monitoring associated with fiction
2. Monitoring of oil conditions

In order to categorize lubricants under various operating conditions, machine learning techniques like artificial neural networks (ANNs) and linear discriminant analysis (LDA) have been widely used.

7.2.2 Forecasting the Behavior of the Tribological System

The combination of information technology and tribology is represented by the prediction of tribological behavior (Figure 7.3). These forecasts are essential for predicting mechanical system failures and determining how long they will last. Cutting processes [7] and Friction stir welding are the applications of predictive tribology, where the addition and removal of materials are the important factors.

Following is a discussion of the three main categories of prediction.

7.2.2.1 Predictive Wear

Wear prediction is a key topic in tribological research since friction-induced wear is the main factor causing component damage. There are several categories for wear prediction as explained next.

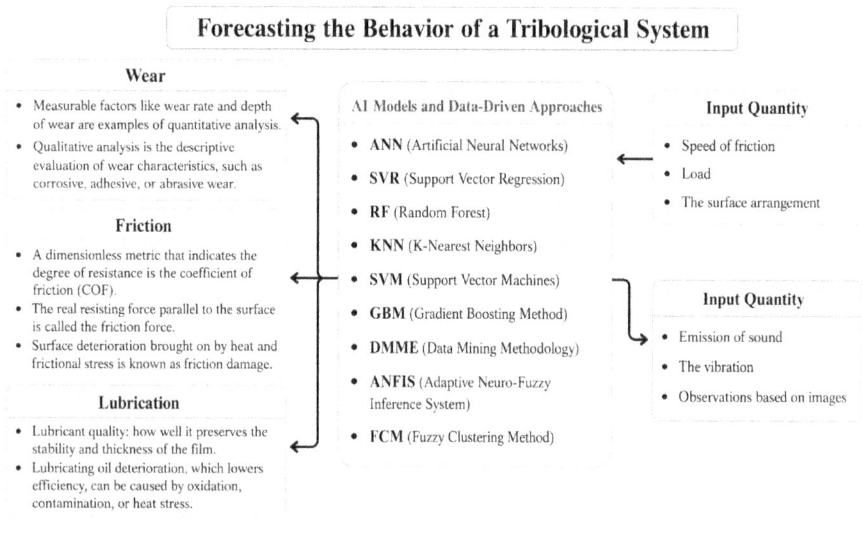

FIGURE 7.3 Forecasting the behavior of a tribological system.

7.2.2.1.1 Wear Quantitative Analysis

In order to do this, it is necessary to analyze large datasets and correlate them with operating time in order to predict parameters like wear rate and wear volume. Among the methods employed are:

- Artificial neural networks (ANNs)
- Random forest (RF)
- Support vector regression (SVR)
- Data mining methodology for engineering (DMME) [8]

Typically, machine learning techniques employ fuzzy inference systems to embed physical rules, neural networks for data learning, and genetic programming to derive mathematical expressions for wear [9].

The majority of research focuses on tool wear [10], and wear rates are predicted by processing large experimental datasets. Other studies use machine learning techniques to examine how surface textures, biological friction materials, and composite materials affect wear.

Several algorithms, including k-nearest neighbor (KNN), SVM, ANN, RF, and gradient boosting machine (GBM), were compared by Hasan et al. [11]. The findings indicated that while RF was more successful in predicting wear, GBM was better at predicting tribological behavior.

7.2.2.1.2 Wear Behavior Classification

A number of variables, including processing condition, material composition, lubrication state, and environmental factors, affect wear behavior, which is qualitative. Among the predictive techniques are:

- Algorithms for clustering
- ANN-based models
- Fuzzy inference
- Adaptive neuro-fuzzy inference system (ANFIS)
- Fuzzy clustering methods (FCM) [12]

Hybrid models that combine ANN, belief rule-based (BRB) inference, and evidential reasoning (ER) models have been proposed for increased reliability because wear classification must also take physical rules into account [13].

7.2.2.2 Friction Forecast

Information technology makes it possible to predict cutting forces in machining, which are impacted by tool geometry, wear, and cutting parameters. Thus, condition monitoring systems and fault prediction are closely related.

- Deep neural networks (DNNs) and SVR are used to create intelligent prediction models that process inputs from monitoring systems.

- The assessment of frictional damage and machining operation also makes use of friction prediction.

Predicting the coefficient of friction (COF) in applications like pipe flows, surface coating, and automotive components is an important field of study. In addition to predicting future system performance, a precise forecast guides material selection and surface treatment procedures, making COF an essential performance metric in tribology. In this area, DNNs work especially well [14].

7.2.2.3 Forecast for Lubrication

The goal of lubrication prediction is to assess and predict lubricant performance. This comprises:

- The connection between lubrication and surface properties (e.g., texture toughness) [15, 16]
- The prediction of lubricant service life [17]
- The change in lubricant characteristics over time [18]

Tribological system optimization and operational life extension depend on accurate lubrication prediction.

7.3 SYSTEM OPTIMIZATION APPROACHES

7.3.1 Tribological System Optimization

Enhancing a tribological system's operational performance through the use of tribological behavior predictions is known as optimization (Figure 7.4). Tribo-pairs, lubricants, and operating conditions make up a tribological system, so all three must be taken into consideration during optimization.

In this field, artificial intelligence (AI) and machine learning (ML) techniques have shown great promise by increasing design efficiency and lowering related expenses, but it is crucial to understand that tribology systems optimize a methodical engineering problem that calls for an all-encompassing strategy that takes into account lubricants' working conditions and tribo-pairs at the same time.

7.3.1.1 Tribo-Pair Optimization

Pre-research avenues can be used to optimize tribo-pairs material design, surface texture design, and material selection.

1. Optimization of surface texture
   It is possible to measure friction behavior experimentally under various surface texture conditions and then use AI techniques to assess how the manufacturing process affects the friction. This makes it possible to create surface texture with parameters that are specially designed to regulate frictional behavior [19, 20]. Tribological system optimization occurs when

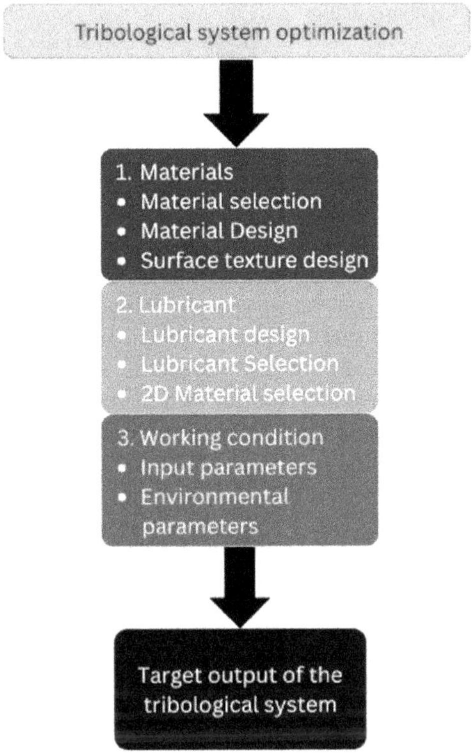

FIGURE 7.4 Tribological system optimization.

the surface texture or preparation procedure is changed backward from a desired friction state [21, 22].

2. Optimization of surface material
   Material development can also be a focus of optimization for intense artificial neural networks. It can be used to efficiently optimize process parameters of treatments like composition of a material and its structure, the latter of which is mostly determined by process parameters, which are the two adjustable components of material design [23].
3. Selection of surface materials
   In real-world engineering applications, choosing the right material is frequently easier than creating brand new ones. To find the materials that are most appropriate for a particular task, a sizable collection of candidate materials must be screened. This is a process facilitated by information technology. For example, Luo et al. [24] created a database of industrial tribological coatings, evaluating coating performance using simple tests before matching it to engineering requirements. High-throughput material screening and selection are made possible by decision tree-based models, which are especially helpful in this situation [25, 26].

7.3.1.2 Lubricant Optimization

Because lubricants are essential for lowering wear and friction, tribological system optimization depends on their selection and design. The main categories of research in this field are new lubricant synthesis [27, 28] and lubricant composition entry shop optimization [29, 30].

The development of two-dimensional (2D) materials has made their application and screening a crucial component of lubricant optimization.

7.3.1.3 Design of Workplace

A tribological system's performance is directly impacted by its input parameters, which include sliding speed and applied load [31, 32], as well as environmental factors, such as temperature and humidity [33, 34]. This working condition must frequently be properly designed and adjusted in order to produce the intended operational outputs. By doing this, a completely optimized tribological system can be realized.

7.4 INTEGRATION WITH IOT AND SMART MANUFACTURING

7.4.1 An Overview of IoT in the Manufacturing System

Numerous industries, including manufacturing, healthcare, transportation, smart cities, and smart homes, have adopted the Internet of Things (IoT) with emphasis on the manufacturing execution system (MES), sensor-based modeling and control, and new advertising in IoT-enabled applications. This section examines the role of IoT in the manufacturing sector.

Shop floor operations and upper-level management tasks are connected by the MES. Establishing transparent data sharing and smooth communication between machines, controllers, and management departments in manufacturing facilities is its primary goal.

Various proprietary systems, including CNC controllers, PLCs, and wireless sensor networks (WSNs), gather operational data at the process level. These real-time data streams are sent to central database servers via gateway computers. Software programs are then used by managers for analysis, monitoring, and decision-making.

MES provides the digital framework for a number of crucial functions:

- Cost analysis, quality control performance, and energy management
- Optimization of the supply chain

MES has changed dramatically as a result of the adoption of IoT technologies. Cloud-based IoT-enabled protocols like MT Connect have made platforms the new standard for MES. Cloud MES make communication, data storage, analytics, and reporting simpler by resolving compatibility issues with proprietary systems.

Enterprise resource planning (ERP) and a process control system (PCS) are integrated with modern practice. Bidirectional data flow:

- Top-down (ERP to PCS): ERP creates purchase orders and work instructions for MES after processing customer orders, market trends, and forecasts. These are further divided by MES into comprehensive production plans that cover machine settings, operator scheduling, and resource allocation.
- Bottom-up (PCS to ERP): PCS gathers in-process information like job status, machine conditions, and sensor readings. For process modifications like predictive maintenance or shift rescheduling, this data is returned to MES. Additionally, MES gives ERP information on labor efficiency, material requirements, and asset utilization, which enhances forecasting, inventory control, and cost analysis.

7.4.2 Sensor-Based Control and Informatics for Manufacturing

A large amount of data is produced by sensing technology in ERP, MES, and PCS. A large portion of this data is still underutilized. New bridges are required to support smart manufacturing and extract significant patterns to fully utilize it. Four domains can be used to group this field of study:

1. Data representation and visualization
   Efficient data representation is crucial due to energy and bandwidth limitations, especially for wireless sensors and RFID tags. Condensed, useful coefficients are created from raw sensor signals by employing techniques such as wavelet and Fourier analysis. This not only reduces storage requirements but also uncovers hidden patterns. With the use of visualization tools like interactive dashboards and network maps, managers can also keep an eye on energy usage, system connections, and key performance indicators (KPIs).
2. Fusion of sensor data
   In a manufacturing system, multiple scenes are commonly used to monitor different aspects of the same process. Sensor fusion provides a thorough grasp of obsession dynamics by merging the different signals. Advanced methods such as recurrence analysis and phase synchronization are employed to examine nonlinear dependence and born linear correlation. In industries like chemical-mechanical processes and the auto industry, these techniques have been applied to promote monitoring accuracy and predictive power.
3. Process control and decision-making
   Following the identification of abnormalities, corrective action is the next step. Statistical models and physics-based failure analysis are examples of conventional techniques. However, this is often limited in its accuracy. Data-driven techniques that make use of real-time data, such as search data, provide more precise prognostics and root-cause identification. Decision support is further improved by discrete event simulation (DES). This model is computationally intensive and tracks individual entities. Continuous flow

models, which depict the process as fluid dynamics, enable faster and more accurate predictions for large-scale manufacturing.

7.4.3 IoT Use in Manufacturing

IoT has recently led to the rise of new paradigms like cloud manufacturing, cyber-physical systems, and industrial IoT. Among the key areas of application are:

1. Cloud manufacturing based on IoT
   By combining WSNs and cloud computing, you can improve real-time monitoring, predictive maintenance, inventory management, and traceability. Cloud analytics has been used to improve productivity and decision-making through the development of architectures like AMCoT and CCIoT–CMfg.
2. Cyber physical manufacturing system (CPMS)
   Predictive maintenance, adaptive production scheduling, and resource task optimization are made easier by integrating virtual IT-based systems with physical material-based systems. Frameworks like IIHub and UML4IoT facilitate the integration of both new and legacy systems into IoT environments.
3. Energy efficiency management
   To reduce emissions and energy consumption, an IoT device streamlines the process and tracks energy use in real time. Examples include additive manufacturing and selective laser sintering, where IoT-based monitoring increases productivity and decreases waste.
4. Management of operations
   Distributed sensing, dynamic scheduling, predictive maintenance, and enhanced quality control are all made possible by IoT. Future, intense, just-in-time scheduling frameworks optimize resource allocation in real time, and IoT-based systems in coal mines identify machine anomalies.
5. Ergonomics and safety
   Risks to worker safety, location, and health are tracked by IoT. In sectors like construction and pharmaceuticals, systems have been developed to monitor physiological status, track unsafe behavior, and enhance accountability.
6. Logistics and supply chain
   Concepts like the physical Internet apply to IoT resource matching, decentralized warehouses, logistics synchronization, and RFID-based inventory management. Global supply chains are made more efficient, traceable, and adaptable by IoT-based platforms.

7.5 CHALLENGES AND LIMITATIONS

The efficient identification of relationships between tribology quantities by informatics techniques makes it possible to monitor states, predictive behavior, and optimize systems. Industrial tribology, lubrication, bio-tribology, friction processing, and geo-tribology are just a few of the fields in which these techniques have since been used.

Analytic techniques can:

- Forecast the service life of tribology systems
- Continually monitor system health
- Predict the service life of the tribology system by analyzing the signal generated from friction, acoustic, image, vibration, thermal, and electrical data
- Find novel connections between tribology variables

However, it is crucial to remember that tribo-informatics is a comprehensive and living field of study rather than just the application of informatics techniques to discrete tribological issues. Even with great advancement, there are still some restrictions on existing applications.

7.5.1 Tribological System Models Are Not Given Enough Consideration

Tribological system models are not sufficiently included in the majority of applications. It is challenging to accomplish profound fusion between informatics and tribology because of the lack of integration. The behavior of an inter-system cannot be adequately captured by the relationship between just two physical variables because tribology is by its very nature systematic, multidisciplinary, and time dependent.

Tribo-informatics has been proposed as a concept and framework to address. Informatics techniques in real-world applications should take into consideration the tribological system inputs, outputs, and state variables, in addition to concentrating on a mover through examination of its underlying principles.

7.5.2 The Challenge of Choosing Appropriate Informatics Techniques

Finding the best informatics approach for a given tribological issue presents another difficulty. Currently, the standard procedure entails choosing a monitoring or prediction method and then comparing the outcomes with experimental data.

However, the tribological problem being studied determines the applicability and efficacy of various informatics techniques. Determining how to apply this technique while maintaining accuracy and efficiency is, therefore, a crucial area of research. This will make it possible for researchers to more effectively adapt informatics techniques to particular tribological platforms.

7.6 FUTURE RESEARCH DIRECTIONS

Tribo-informatics will continue to grow in scope and significance as information technology continues to progress. The effective application of techniques like regression, classification, clustering, and dimensionality reduction is hampered by several issues. These difficulties fall into the following categories.

7.6.1 Absence of a True Framework of Tribo-Information

Currently, there isn't a clear framework that combines informatics tribology completely. Tribo-informatics is still used in research in a rather haphazard and disjointed manner. For example, we are seeing far more research in friction monitoring than in condition monitoring of tribology systems.

A systematic concept and framework must be established to fortify tribo-informatics and expand its application. Once created, the search framework can direct extensive research that uses it as an information-based tribology model to connect the inputs, states, and outputs of tribological systems. More precise steam behavior monitoring, forecasting, and optimization will be made possible by this integration.

7.6.2 Standardized Tribological Examinations Are Necessary

Big datasets are essential to tribo-informatics. However, the value of nonstandardized tribological tests for wider application is limited because they frequently produce inconsistent or useless data. To guarantee that tribological data are trustworthy, reusable, and broadly applicable, a standard testing protocol must be established.

The resulting dataset should not be discarded after a single study. For instance, if a researcher uses a standardized pin-on-disk test to examine the relationship between the wear scar shape and friction, alternatively, the connection between vibration, acoustic signals, and frictionless state in conjunction with findings from other standardized pin-on-disk tests could be investigated. By enhancing the reusability, circulation, and dissemination of tribological data, these protocols will increase the effectiveness and impact of tribological research.

7.6.3 Technological Advancement in the Gathering of Tribological Data

The foundation of tribo-informatics is the collection of reliable data. Therefore, advanced sensing technologies are needed to collect sufficient and accurate tribological data. Due to the fact that electrical signals are among the methods available, they are especially attractive because they are comparatively simple to measure and analyze. It is projected that new technologies such as self-powered triboelectric nanogenerators (TENGs) [36,37] and intelligent sensing coatings for active tribology [35] will play a significant role in data collection in the future. Deep self-powered sensors can offer a very effective and practical way to gather tribological data once the effect of variables like temperature, humidity, vibration frequency, and material characteristics on triboelectric signals is carefully separated.

Engineers in a variety of fields can view surfaces and structures in novel ways with the aid of coating technologies. These consist of surface roughness testing, mechanical integrity tests, such as tensile, compressive, and flexural strength tests, and enhancements to solar energy systems' performance through modifications to their optical and thermal characteristics. Additionally, coatings make it easier to explain how fluids flow through difficult-to-understand networks of channels and transitions (such as pipes and elbows). Functional coatings facilitate the prediction of the behavior and effects of loads on precision-engineered systems, such as ergonomically sensitive seating systems. This improves the design, extends the life of the systems, and reduces errors in multiphysics simulations [38–55].

7.7 SUMMARY

This chapter emphasizes how data analytics, artificial intelligence (AI), and machine learning (ML) are increasingly contributing to the advancement of tribology behind the conventional approach to the study of wear, friction, and lubrication. Condition monitoring, behavior prediction, and system optimization are the three main areas it highlights. Condition monitoring uses machine learning models to identify wear, friction states, and lubrication quality in real time using vibration, acoustic, image, and oil analytics data. In order to forecast wear progression, friction dynamics, and lubricant degradation, and increase component life and reliability, predictive modeling employs algorithms like ANN, SVM, and random forest. In order to improve performance and cut costs, optimization strategies use AI to design materials, apply textures, create lubricants, and modify working conditions. The integration of IoT and smart manufacturing are also examined, where supply chain optimization, energy efficiency, and predictive maintenance are made possible by sensor-based monitoring, cloud platforms, and cyber-physical systems. Notwithstanding notable advancements, issues with model integration, method selection, and the absence of standardized datasets still exist. To increase the data application and reliability, future studies should concentrate on standardized testing procedures, unified tribo-informatics frameworks, and sophisticated sensing technologies. When all is said and done, data-driven approaches have the power to revolutionize tribology and enhance the efficiency, predictability, and sustainability of engineering systems.

REFERENCES

1. Patil S, Phalle V. Fault detection of anti-friction bearing using ensemble machine learning methods. *Int J Eng* 31(11): 1972–1981 (2018).
2. Leandro Cabral MA, Matamoros EP, Ferreira Costa JA, Vieira Pinto AP, Silva de Souza A, Freire AD, Filgueira Bezerra CE, Santos Cabral ELD, Silva Castro WR, Pires de Souza R, et al. Automated classification of tribological faults of alternative systems with the use of unsupervised artificial neural networks. *J Comput Theor Nanosci* 16(7): 2644–2659 (2019).
3. Kilundu B, Dehombreux P, Chiementin X. Tool wear monitoring by machine learning techniques and singular spectrum analysis. *Mech Syst Signal Process* 25(1): 400–415 (2011).
4. Chang HC, Borghesani P, Peng ZX. Automated assessment of gear wear mechanism and severity using mould images and convolutional neural networks. *Tribol Int* 147: 106280 (2020).
5. Yan XP, Zhao CH, Lu ZY, Zhou XC, Xiao HL. A study of information technology used in oil monitoring. *Tribol Int* 38(10): 879–886 (2005).
6. Yuan W, Chin KS, Hua M, Dong GN, Wang CH. Shape classification of wear particles by image boundary analysis using machine learning algorithms. *Mech Syst Signal Process* 72–73: 346–358 (2016).

7. Moder J, Bergmann P, Grün F. Lubrication regime classification of hydrodynamic journal bearings by machine learning using torque data. *Lubricants* 6(4): 108 (2018).
8. Farias A, Almeida SLR, Delijaicov S, Seriacopi V, Bordinassi E C. Simple machine learning allied with data-driven methods for monitoring tool wear in machining processes. *Int J Adv Manuf Technol* 109(9–12): 2491–2501 (2020).
9. Kolodziejczyk T, Toscano R, Fouvry S, Morales-Espejel G. Artificial intelligence as an efficient technique for ball bearing fretting wear damage prediction. *Wear* 268(1–2): 309–315 (2010).
10. Bellotti M, Wu M, Qian J, Reynaerts D. Tool wear and material removal predictions in micro-EDM drilling: Advantages of data-driven approaches. *Appl Sci* 10(18): 6357 (2020).
11. Hasan MS, Kordijazi A, Rohatgi PK, Nosonovsky M. Triboinformatic modeling of dry friction and wear of aluminum base alloys using machine learning algorithms. *Tribol Int* 161: 107065 (2021).
12. Alambeigi F, Khadem SM, Khorsand H, Mirza Seied Hasan E. A comparison of the performance of artificial intelligence methods in the prediction of dry sliding wear behavior. *Int J Adv Manuf Technol* 84(9–12): 1981–1994 (2016).
13. Xu XJ, Zhao ZZ, Xu XB, Yang JB, Chang LL, Yan XP, Wang GD. Machine learning-based wear fault diagnosis for marine diesel engine by fusing multiple data-driven models. *Knowl Based Syst* 190: 105324 (2020).
14. Šabanovič E, Žuraulis V, Prentkovskis O, Skrickij V. Identification of road-surface type using deep neural networks for friction coefficient estimation. *Sensors* 20(3): 612 (2020).
15. Gropper D, Wang L, Harvey TJ. Hydrodynamic lubrication of textured surfaces: A review of modeling techniques and key findings. *Tribol Int* 94: 509–529 (2016).
16. Kurdi A, Alhazmi N, Alhazmi H, Tabbakh T. Practice of simulation and life cycle assessment in tribology—A review. *Materials* 13(16): 3489 (2020).
17. Zhang Y, Xu XJ. Machine learning decomposition onset temperature of lubricant additives. *J Mater Eng Perform* 29(10): 6605–6616 (2020).
18. Desai PS, Granja V, Higgs CF III. Lifetime prediction using a tribology-aware, deep learning-based digital twin of ball bearing-like tribosystems in oil and gas. *Processes* 9(6): 922 (2021).
19. Zambrano V, Brase M, Hernández-Gascón B, Wangenheim M, Gracia LA, Viejo I, Izquierdo S, Valdés JR. A digital twin for friction prediction in dynamic rubber applications with surface textures. *Lubricants* 9(5): 57 (2021).
20. Kordijazi A, Behera S, Patel D, Rohatgi P, Nosonovsky M. Predictive analysis of wettability of Al–Si based multiphase alloys and aluminum matrix composites by machine learning and physical modeling. *Langmuir* 37(12): 3766–3777 (2021).
21. Grützmacher PG, Profito FJ, Rosenkranz A. Multi-scale surface texturing in tribology—Current knowledge and future perspectives. *Lubricants* 7(11): 95 (2019).
22. Cavaleri L, Asteris PG, Psyllaki PP, Douvika MG, Skentou A D, Vaxevanidis NM. Prediction of surface treatment effects on the tribological performance of tool steels using artificial neural networks. *Appl Sci* 9(14): 2788 (2019).
23. Kordijazi A, Roshan HM, Dhingra A, Povolo M, Rohatgi PK, Nosonovsky M. Machine-learning methods to predict the wetting properties of iron-based composites. *Surf Innov* 9(2–3): 111–119 (2021).
24. Luo DB, Fridrici V, Kapsa P. A systematic approach for the selection of tribological coatings. *Wear* 271(9–10): 2132–2143 (2011).
25. Bucholz EW, Kong CS, Marchman KR, Sawyer WG, Phillpot SR, Sinnott SB, Rajan K. Data-driven model for estimation of friction coefficient via informatics methods. *Tribol Lett* 47(2): 211–221 (2012).

26. Tysoe WT, Spencer ND. Designing lubricants by artificial intelligence. *Tribol Lubr Technol* 76(6): 78 (2020).
27. Bhaumik S, Pathak SD, Dey S, Datta S. Artificial intelligence-based design of multiple friction modifiers dispersed castor oil and evaluating its tribological properties. *Tribol Int* 140: 105813 (2019).
28. Jia D, Duan H, Zhan S, Jin Y, Cheng B, Li J. Design and development of lubricating material database and research on performance prediction method of machine learning. *Sci Reports* 9: 20277 (2019).
29. Bhaumik S, Kamaraj M. Artificial neural network and multi-criterion decision making approach of designing a blend of biodegradable lubricants and investigating its tribological properties. *Proc Inst Mech Eng Part J J Eng Tribol* 235(8): 1575–1589 (2021).
30. Humelnicu, Ciortan, Amortila. Artificial neural network-based analysis of the tribological behavior of vegetable oil–diesel fuel mixtures. *Lubricants* 7(4): 32 (2019).
31. Ramkumar T, Selvakumar M. Artificial intelligence in predicting the optimized wear behaviour parameters of sintered titanium grade 5 reinforced with nano B4C particles. In: *Artificial Intelligence in Mechanical and Industrial Engineering*. Boca Raton: CRC Press, 2021.
32. Babin A. Data-driven system identification and optimal control of an active rotor-bearing system. *IOP Conf Ser: Mater Sci Eng* 1047(1): 012053 (2021).
33. Wierzcholski K, Łupicka O. Network analysis for artificial intelligence micro-bearing systems. *J KONES* 16(3): 489–495 (2009).
34. Li JY, Yao XX, Zhang Z. Physical model based on data-driven analysis of chemical composition effects of friction stir welding. *J Mater Eng Perform* 29(10): 6591–6604 (2020).
35. Zhang ZN, Yin N, Wu ZS, Pan SH, Wang DA. Research methods of contact electrification: Theoretical simulation and experiment. *Nano Energy* 79: 105501 (2021).
36. Zhou LL, Liu D, Wang J, Wang ZL. Triboelectric nano generators: Fundamental physics and potential applications. *Friction* 8(3): 481–506 (2020).
37. Wang HD, Ma GZ, Xu BS, Yong QS, He PF. Design and application of friction pair surface modification coating for remanufacturing. *Friction* 5(3): 351–360 (2017).
38. Verma PK, Vasudev H, Kumar V. Dental implant materials and designs and failure during implant placement and loading: A review. In: *Advanced Manufacturing Processes*. CRC Press, 2025: 138–150.
39. Verma PK, Vasudev H, Kumar V. Bioactivity and corrosion performance of low-velocity oxy-fuel-sprayed hydroxyapatite-based coatings on a surgical-grade stainless steel. *Trans Indian Inst Met* 78(8): 185 (2025).
40. Verma PK, Vasudev H, Kumar V. Effect of electro-corrosion behavior of different grades of stainless steel used in bio-implants: a review. In: *Advanced Manufacturing Processes*. CRC Press, 2025: 269–283.
41. Verma, PK, Vasudev H. Importance of hydroxyapatite coatings in improving new bone osteogenesis: A review. In: *Advanced Manufacturing Processes*. CRC Press, 2025: 284–299.
42. Dutta V, Thakur L, Singh B, Vasudev H. A study of erosion–corrosion behaviour of friction stir-processed chromium-reinforced NiAl bronze composite. *Materials* 15(15): 5401 (2022).
43. Singh M, Vasudev H, Kumar R. Corrosion and tribological behaviour of BN thin films deposited using magnetron sputtering. *Int J Surf Eng Interdiscip Mater Sci* 9(2): 24–39 (2021).
44. Sharanabasava, H, Raviprakash M, Prasad CD, Ramesh M R, Phanibhushana MV, Vasudev H, Kumar S. Microstructure, mechanical and wear properties of SiC and Mo reinforced NiCr microwave cladding. *Adv Mater Process Technol* 10: 3620–3633 (2023).

45. Singh M, Vasudev H, Singh M. Surface protection of SS-316L with boron nitride based thin films using radio frequency magnetron sputtering technique. *J Electrochem Sci Eng* 12(5): 851–863 (2022).
46. Prashar G, Vasudev H, Thakur L, Bansal A. Performance of thermally sprayed hydroxyapatite coatings for biomedical implants: a comprehensive review. *Surf Rev Lett* 30(01): 2241001 (2023).
47. Vasudev H. Wear characteristics of Ni-WC powder deposited by using a microwave route on mild steel: Microwave cladding of Ni-WC. *Int J Surf Eng Interdiscip Mater Sci* 8(1): 44–54 (2020).
48. Parkash J, Saggu H S, Vasudev H. A short review on the performance of high velocity oxy-fuel coatings in boiler steel applications. *Mater Today Proc* 50: 1442–1446 (2022).
49. Singh P, Vasudev H, Bansal A. Effect of post-heat treatment on the microstructural, mechanical, and bioactivity behavior of the microwave-assisted alumina-reinforced hydroxyapatite cladding. *Proc Inst Mech Eng Part E J Process Mech Eng* 238(3): 1060–1071 (2024).
50. Verma, PK, Vasudev H, Singh G, Singh S. A review on the microstructural and biomedical properties of implants manufactured using additive manufacturing. *J Inst Eng India Ser D* 106: 1–15 (2024).
51. Singh SP, Verma PK, Singh S, Kapoor M, Balu R. Microstructural characterization and electrochemical behaviour of HVOF sprayed WC-Co-Cr coatings on 316L boiler steel stainless steel. *Results Surfaces Interfaces* 16: 100248 (2024).
52. Verma PK, Minhas AS, Singh P, Kumar S. Slurry erosion behaviour of thermal sprayed Al2O3 and Cr2O3 coatings for turbine steels. *AIP Conf Proc* 2986(1): 020012 (2024).
53. Verma P K, Kumar V, Vasudev H. Bioactivity and corrosion analysis of thermally sprayed hydroxyapatite based coatings. *J Electrochem Sci Eng* 14(4): 419–439 (2024).
54. Verma P K, Vasudev H, Kumar V. Biocompatibility and corrosion performance of HA-based LVOF sprayed and post-annealed coatings on SS 254 surgical-grade bio-implant material. *Ceram Int* 51(10): 13207–13225 (2025).
55. Verma PK, Vasudev H, Singh G. Improving the corrosion performance of SS 304 stainless steel using HVOF-sprayed WC-Co-Cr coatings in chloride environments. *J Bio-Tribo-Corros* 11(4): 1–13 (2025).

8 Addressing Challenges and Exploring Opportunities in Implementing a Smart Tribology System

Sanjay W. Rukhande, Shubhangi Suryawanshi, Digvijay Bhosale, Satyajeet Gaikwad, and Gursimran Kaur

8.1 INTRODUCTION

The science of friction, wear, and lubrication known as tribology is key in determining the reliability and the longevity of a mechanical system. Its industrial value cuts across industries ranging between aerospace and manufacturing whereby performance, energy performance, and maintenance costs are directly affected by efficient surface interactions [1, 2]. The application of smart tribology is a paradigm shift in wear and friction studies where the traditional wear and friction research is substituted by data-driven sensor-bound systems capable of real-time performance evaluation. This dynamic discipline presents enormous opportunities on how to improve reliability, efficiency, and predictive maintenance in a wide range of applications of engineering. Smart tribology systems are increasing in capabilities of real-time sensing, data analytics, and actuation due to the progress of low-cost sensor technologies and Industry 4.0 connectivity to monitor the responsive and adaptive performance of surfaces [3]. This integration of tribology and digital intelligence, also known as tribotronics, provides new avenues of condition-based maintenance and intelligent surface control, but introduces implementation issues such as sensor integration, data processing, and system reliability [4]. The rise in awareness of the effect of tribology that has contributed to almost a quarter of the world energy consumption explains the importance of the need to understand intelligent surface systems that would control wear and friction in real time. However, implementation of smart tribology in the real world encounters challenges of robustness of sensors, generalization of the

DOI: 10.1201/9781003741152-8

analytical model, and non-disruptive integration with surface engineering methods [5]. This chapter will delimit the key implementation goals of a smart tribology system like real-time tribological monitoring, predictive wear analytics, and adaptive feedback control.

8.2 BACKGROUND AND LITERATURE REVIEW

Conventional tribology has been concerned with the examination of friction, wear, and lubrication over the years through experimental measurements and physics modeling of the contacting surfaces to guarantee dependability and lifespan in mechanical systems. Conversely, smart tribology uses real-time sensing, data-driven analytics, and AI-refined feedback solutions so that it can support predictive wear-based and adaptive surface protection solutions [6, 7]. The recent developments in tribology have seen the adoption of a wide range of sensor technologies, including friction, wear, vibration, and acoustic emission sensors, to analyze data and make predictions and real-time condition monitoring with the help of machine learning algorithms and data analytics. The use of the Internet of Things (IoT) based on architectures and other methods such as digital twins and tribo-informatics has also contributed toward the ability of smart tribology systems to optimize maintenance plans and surface performance results [6, 8]. More recent developments in smart tribology have used data-driven strategies (e.g., deep learning and digital twins) to predict wear and frictional behavior in complex tribosystems, with much greater predictive maintenance performance than conventional strategies [9]. The art of thermal sprayed functional and smart coatings, such as those deposited through Ni-based and other alloy systems, has been broadened with a number of novel capabilities, including self-healing, sensor integration, and adaptive responses to environmental stimuli [4]. A relatively new and effective tool in tribo-informatics is machine learning, which allows predicting the friction and wear on textured and porous surfaces with high accuracy to open new opportunities in the context of introducing intelligent analytics into tribological surfaces design [10]. Although there has been a remarkable advancement in the field of sensing technology integration and analytics on data, a majority of the studies conducted in the area center on independent parameters, as opposed to the overall system performance. There has been little research on the joint issues of real-time wear detection, oxidation monitoring, and predictive modeling of Ni-based coatings at service-like conditions. Also, the shift to field-ready and robust smart tribology systems is not thoroughly investigated so far, and the gap between the experimental demonstrations and the industrial applications is frightening.

No specific standard is currently used to implement smart tribology systems, but the ISO 17359:2018 standard is aimed at general guidelines of condition monitoring and diagnostics of machine systems including tribology parameters by proposing structured procedures of data acquisition, alarm settings, diagnosis, and prognosis [11]. Other standards developed in the specialized field include ISO 14830-1:2019 that focuses on tribology monitoring and helping to analyze lubricant and wear debris in a proactive maintenance context. The IoT-compatible communication standards, such as IEEE 1451, can facilitate the sensor interface of smart systems with metadata-rich information and interaction capabilities.

8.3 SMART TRIBOLOGY SYSTEM: CONCEPT AND ARCHITECTURE

A smart tribology system (STS) is a connected system that involves real-time sensors, edge-computing sensors, and data analytics to observe, predict, and proactively control friction,wear, and surface wear in mechanical contacts (Figure 8.1) [12]. An STS consists of major parts such as embedded or contact-based sensors of measurements, including acoustic emission, temperature, and vibration; an acquisition and processing unit (usually with local intelligence) to extract features and detect anomalies; and a communication layer to generate alerts or feedback control to maintain

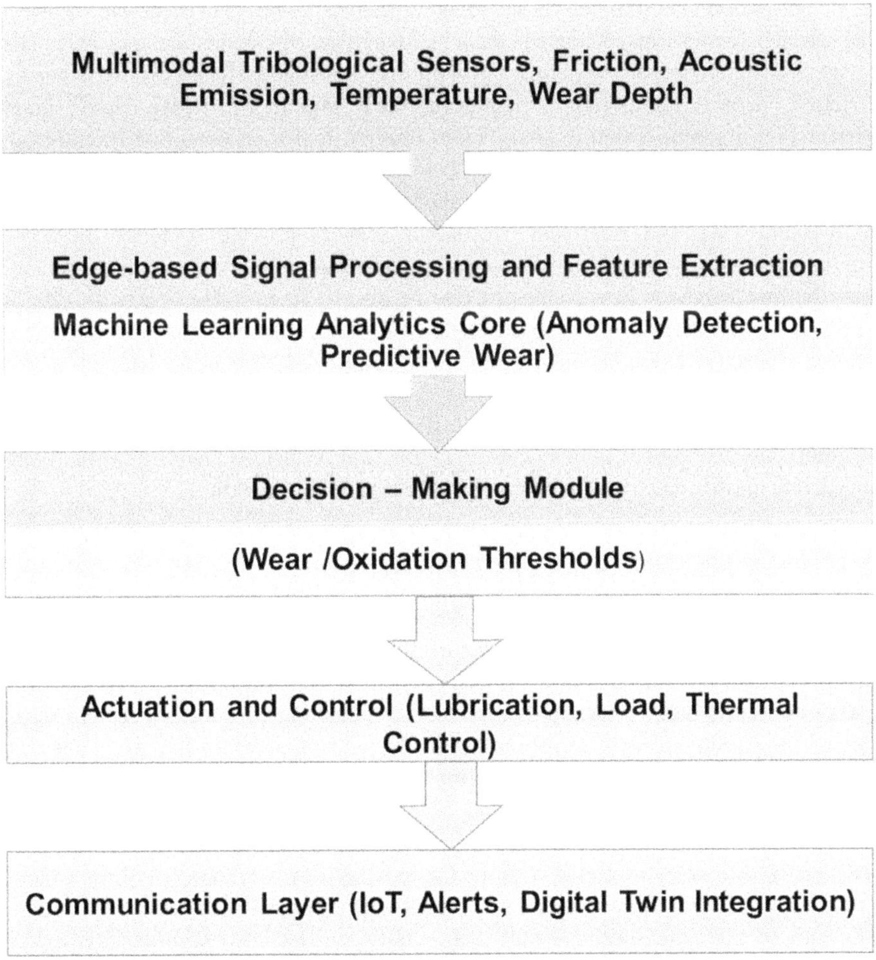

FIGURE 8.1 Smart tribology system block diagram.

or adapt behavior [3]. Through predictive analytics and related intelligence, STS is no longer passive in measurements, but proactive in its decisions, predictive in maintenance, and adaptive in tribological control strategies that promote better performance of the coating and reliability of the system. Tomasz Liskiewicz et al. [3] present an overview of the recent sensor technologies used to monitor tribological parameters in real-time, including the role of integrated sensing and data capture in developing tribotronics and intelligent condition monitoring systems.

The architecture combines multisensor tribological data pipelines of friction, wear, acoustic emission, and temperature with a machine learning analytics core (based on supervised and unsupervised models) to conduct real-time condition assessment, anomaly detection, and predictive wear. Under the architecture of the smart tribology system, the real-time sensor data (e.g., friction, acoustic emission, temperature) is processed by a decision-making module to activate a desired actuation (e.g., lubrication injection, load adjustment, or thermal control) to proactively protect the surface (waveforms) when the wear or oxidation thresholds are met [12]. The suggested smart tribology system design combines multimodal tribological measuring (friction, acoustic emission, temperature, wear depth) with edge-based signal processing, machine-learning analytics to provide real-time condition monitoring, predictive wear forecast, and closed-loop management of surface performance [3]. The materials-integrated sensors like embedded piezoresistive coating, luminescent multilayer films, and microwave resonator sensors provide some of the most sophisticated ways of tracking tribological interface degradation, wear development, and thermal behavior in a nondestructive manner [13]. Erik Hansen et al. (2023) [14] established the ability of a digital twin of a pin-on-disk tribometer to accurately predict physical friction behavior using calibrated mixed lubrication theories, which is a strong base of real-time simulation-controlled control. This smart tribology system is harmoniously built into the Industry 4.0 models as it is a blend of IoT-based sensing, two-way digital twin synchronization, and AI-driven analytics to facilitate predictive maintenance and autonomous tribological optimization.

8.4 CHALLENGES IN IMPLEMENTATION

There are major challenges in embedding durable sensors directly onto tribologically active surfaces, such as keeping the signal integrity when subjected to heavy wear, using thermal cycling, and in corrosive environments. Kiesling et al. (2024) [15] elaborate on the challenging manufacturing and contacting challenges that were faced during the development of thin-film sensors in bearing surfaces that are subjected to mechanical and thermal loads. Sensing in difficult mixed-friction environments (over long periods) is complex because of sensor degradation caused by wear and also because of the requirements in protecting nonobtrusive encapsulation schemes. The technical issues associated with the implementation of a smart tribology system are among the most fundamental, namely, how to guarantee the reliability of data and how to filter measurement noise, especially the low-cost, high-density sensor array to preserve the signal integrity without overwhelming the computational or energy resources. It has proven to be a significant technical challenge to provide real-time responsiveness to a smart tribology system, since latency and bandwidth constraints, particularly in treating high-frequency sensor data at the edge, can defer fault detection and impede timely control measures [6]. The triboinformatic machine-learning model developed by Birada et al. (2024) [16] allows predicting the wear loss in the form of volume and friction of SS316L alloy coated with WC/NiCrBSi and its results are highly accurate, but their work notes that there remains a heavy dependence on the composition of the coating, load, and velocity even with the advanced predictive model, which makes the wear behavior more difficult to generalize. One important issue of critical material concern is compatibility between Ni-based coating and underlying surface treatment since pretreatment technological solutions (e.g., grit blasting or polishing) may profoundly modify the presence of subsurface defects and, thus, influence the oxidation kinetics, adhesion, and tribological life. The various surface preparations of a Ni-based superalloy have a drastic influence on the oxidation kinetics and morphology of oxide scale that are pertinent in coating performance [17]. The initial cost of smart tribology systems, including the advanced sensors, data acquisition hardware, and infrastructure, as well as the difficulties of scaling up the system out of the laboratory to industrial settings, creates a huge impediment to a comprehensive widespread solution. Updating older tribological test rigs with new sensing, data capture, and connection solutions poses an enormous cost- and scale-based challenge, with the complexity of redesigning equipment and integrating them often surpassing the advantages and making them not widely adopted [18].

The application of a smart tribology system is limited by the scarcity of high-quality tribological datasets with labels of failures, as well as the challenge of labeling wear and oxidation events detected by the sensor, which will weaken the creation of reliable AI-based predictive models. The generalization of the model to a wide range of tribological applications is a major challenge because data-driven wear and oxidation predictors that are trained on limited experiments with Ni-based coating do not always predict well when used in new coating experiments or operating regimes [19]. A developing skill mismatch in the workforce also restricts smart tribology deployment and implementation as the current personnel do not have the interdisciplinary data science, sensor engineering, and domain-specific analysis skills to design, deploy, and maintain intelligent tribological systems.

8.5 OPPORTUNITIES AND EMERGING TRENDS

Emerging smart tribology systems that integrate real-time sensor data and machine learning models offer transformative opportunities for predictive maintenance, enabling accurate forecasting of wear and oxidation in Ni-based coatings and proactive intervention strategies. Desai et al. (2021) [9] illustrate how tribological data combined with deep learning and digital-twin frameworks can effectively forecast component lifetime and support maintenance planning. The advent of new AI/ML methodologies, including neural networks and gradient boosting, provides exciting data-driven opportunities for the accurate prediction of friction and the wear and coefficient of friction in tribological systems, which in turn, enables predictive maintenance and improved surface performance in these systems [20]. New developments in smart tribology based on energy-efficient Ni-based coatings and reduced friction through surface engineering provide tremendous sustainable opportunities in reducing tribological energy loss by over ~40% and improving CO_2 emissions through extending the life of parts and reducing resources [21]. Novel nanocomposite and hybrid coating technologies afford possibilities that have never been seen before in terms of tailoring tribological performance (i.e., wear, friction performance, environmental adaptation) via individualization of material composition, microstructure, and functionality. The possible employment of self-healing/adaptive surfaces affords the prospective opportunity to autonomously self-repair microdamage and to dynamically tune tribological characteristics (e.g., through stimulus-triggered healing or tribo-film generation); thereby extending the functional life of Ni-based coatings dramatically in smart tribology systems. The presence of new smart tribology systems provides exciting opportunities for such areas as aerospace, automotive, biomedical, and manufacturing (e.g., in adaptive wear monitoring of high-speed components and intelligent surface diagnostics and monitoring of orthopedic implants and precision tooling) [4].

The tribological performance of nickel–chromium–boron–silicon–iron–carbo n coatings applied by air plasma spraying (APS) and high-velocity oxygen fuel (HVOF) spraying was evaluated using a data-driven approach, enabling predictions of the wear depth and friction coefficient to be made with a high degree of accuracy (Figure 8.2). The experimental technique was performed on a pin-on-disk tribometer (TR-20LE, Ducom Instruments) in accordance with ASTM G99, which was linked with wear depth measurements by a linear variable differential transformer and friction measurement by a rotary friction transducer, and real-time data capture via WINDUCOM 2010 software. For the counter-material, balls of silicon nitride were used because of their hardness and chemical inertness. The coating experiments were designed around five prime parameters: load, sliding velocity, sliding time, sliding distance, and temperature of the tribo-chamber, which were structured by means of a definitive screening design (DSD), enabling the main effect/degree of importance of the various factors to be effectively recognized. The statistical models produced by the regression analysis of the outputs from the DSD give predictions that are within 3.8% and 5.6%, respectively, for the friction coefficient and wear depth, showing the practicality of the models for real-life tribological predictions. The analysis disclosed that the load and the temperature of the chamber were the principal parameters affecting both wear and friction, the best overall results being obtained from the HVOF-produced coatings due to their high hardness and dense microstructure. Scanning electron microscopy (SEM) and energy dispersal x-ray spectrometry (EDS) showed that the surfaces after wear possessed Cr_2O_3, NiO, and SiO_2, which would account for the oxidation protection and the thermal stability. The results give a basis for the incorporation of AI-aided prediction into systems for smart tribology and proactive surface maintenance systems [22, 23].

The correlation observed between the wear parameters resulting from sensor operations and real-life trends in coating degradation demonstrates the potential for integrated monitoring to offer early warning in intelligent tribology. Further to this,

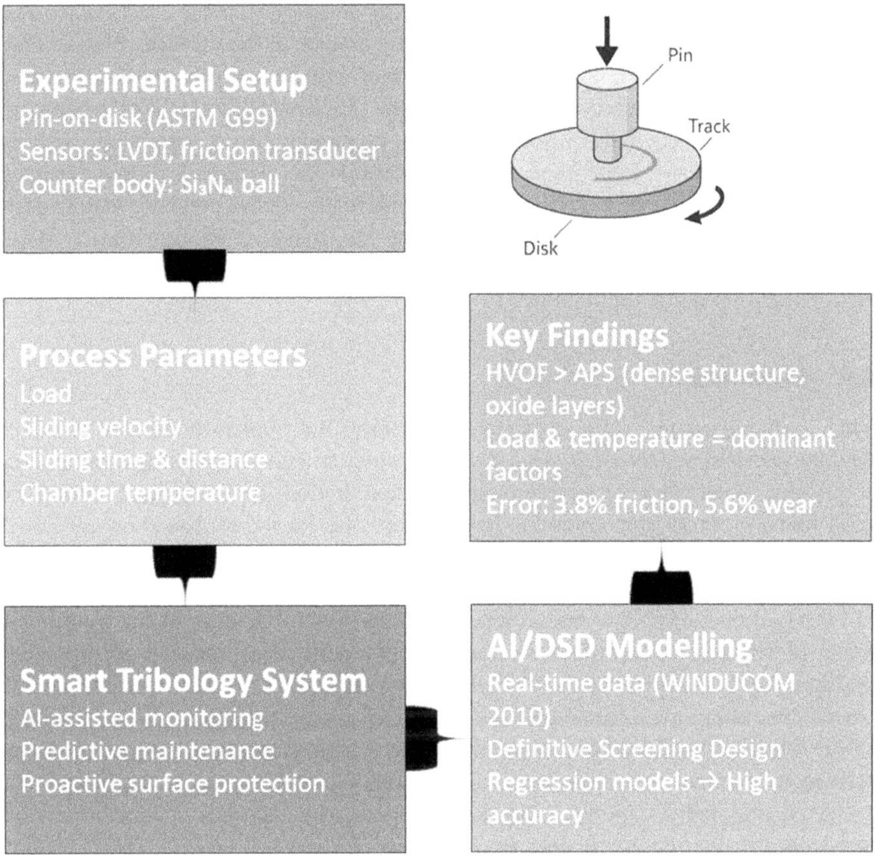

FIGURE 8.2 Integrated experimental and AI modeling workflow for tribological performance.

the advantages of the proposed architecture in its flexibility and predictive capabilities greatly outstrip those offered by conventional tribological systems, especially in respect of the application of the approach to assessment of Ni-based coating performance. The fact that under controlled cyclic loading it has been shown that the sensor–data analysis platform recommended above is technically valid and scalable to real-life applications demonstrates the validity of the findings both in terms of the opportunities for industry to reduce unplanned downtime by predictive maintenance methods and for academia in working upon a structured methodology to enhance intelligent tribological research. From this basis, future efforts should endeavor to combine the multimodal sensing platforms with machine learning approaches based upon a wide variety of coating compositions and environments of applications in order to further enhance the generality of the system.

Coating technologies enable engineers across various disciplines to examine surfaces and structures from innovative perspectives. They encompass assessments of surface roughness, evaluations of mechanical integrity attributes such as tensile, compressive, and flexural strengths, and enhancements in the performance of solar energy systems through modifications of their optical and thermal properties. Coatings also assist in effectively illustrating the movement of fluids within complex networks of channels and transitions that are challenging to comprehend, such as pipelines and elbows. Functional coatings facilitate the prediction of how stresses will impact the structure and how it will perform within precision-engineered systems, such as seating arrangements that are sensitive to ergonomics. This enhances the longevity of the systems, improves the design quality, and reduces errors in multiphysics simulations [24–41].

8.6 SUMMARY

This study raised the main obstacles concerning the implementation of a smart tribology system: sensor integration as implemented in a harsh operating atmosphere, acquisition of data that can be relied upon, and prediction of coating deterioration. In turn, however, promising areas for growth were also identified, based on advanced sensor technology, analytical techniques based on data, and innovations in surface engineering methods directed at maximizing the performance and service lifetime of Ni-based coatings. This research project considers the joining of the traditional areas of tribological testing with the more modern ideas relating to proactive maintenance systems particularly related to Industry 4.0 by presenting experimental verification alongside a framework for intelligent monitoring. It is advised that any future work should focus on, in order of priority, multimodal sensing, techniques relating to adaptive algorithms that can learn relative to the various operational data collected, and interdisciplinary knowledge imparted by materials science, advanced techniques in the realms of data analytics, and mechanical design. These themes should be industry ready to move smart tribology from the level of being a promising construct to a concept that is realistic and achievable to implement.

REFERENCES

1. Johns-Rahnejat, P.M., Rahmani, R. and Rahnejat, H., 2023. Current and future trends in tribological research. *Lubricants*, 11(9), p.391. doi: 10.3390/lubricants11090391.
2. Hutchings, I.M., 1992. Tribology: friction and wear of engineering materials. *Materials & Design*. doi: 10.1016/0261-3069(92)90241-9.
3. Liskiewicz, T., Sherrington, I., Khan, T. and Liu, Y., 2023. Advances in sensing for real-time monitoring of tribological parameters. *Tribology International*, p.108965. doi: 10.1016/j.triboint.2023.108965.
4. Tejero-Martin, D., Rezvani Rad, M., McDonald, A. and Hussain, T., 2019. Beyond traditional coatings: A review on thermal-sprayed functional and smart coatings. *Surface and Coatings Technology*. doi: 10.1007/s11666-019-00857-1.
5. Holmberg, K. and Erdemir, A., 2017. Influence of tribology on global energy consumption, costs and emissions. *Friction*. doi: 10.1007/s40544-017-0183-5.
6. Rouf, S., Raina, A., Ul Haq, M.I. and Naveed, N., 2022. Sensors and tribological systems: applications for Industry 4.0. *Industrial Robot*. doi: 10.1108/IR-10-2021-0225.

7. Yin, N., Yang, P., Liu, S., Pan, S. and Zhang, Z., 2024. AI for tribology: Present and future. *Friction*. doi: 10.1007/s40544-024-0879-2.
8. Yin, N., Xing, Z., He, K. and Zhang, Z., 2023. Tribo-informatics approaches in tribology research: A review. *Friction*. doi: 10.1007/s40544-022-0596-7.
9. Desai, P.S., Granja, V. and Higgs, C.F., 2021. Lifetime prediction using a tribology-aware, deep learning-based digital twin of ball bearing-like tribosystems in oil and gas. *Processes*. doi: 10.3390/pr9060922.
10. Boidi, G., Da Silva, M.R., Profito, F.J. and Machado, I.F., 2020. Using machine learning radial basis function (RBF) method for predicting lubricated friction on textured and porous surfaces. *Surface Topography: Metrology and Properties*. doi: 10.1088/2051-672X/abae13.
11. Rodriguez, S., Granados, D.E. and Suarez, B.C., 2023. Implementation of a condition-based monitoring management system for the pasta preparation area of Trupal S.A. company based on the ISO-17359:2018 standard. In: *Proceedings of the LACCEI International Multi-Conference for Engineering, Education and Technology*. doi: 10.18687/laccei2023.1.1.421.
12. Zhang, Z., Yin, N., Chen, S. and Liu, C., 2021. Tribo-informatics: Concept, architecture, and case study. *Friction*. doi: 10.1007/s40544-020-0457-3.
13. Balasubramanian, V., Niksan, O., Jain, M.C., Golovin, K. and Zarifi, M.H., 2023. Non-destructive erosive wear monitoring of multi-layer coatings using AI-enabled differential split ring resonator based system. *Nature Communications*. doi: 10.1038/s41467-023-40636-9.
14. Hansen, E., Vaitkunaite, G., Schneider, J., Gumbsch, P. and Frohnapfel, B., 2023. Establishment and calibration of a digital twin to replicate the friction behaviour of a pin-on-disk tribometer. *Lubricants*, 11(2), p.75, doi: 10.3390/lubricants11020075.
15. Kiesling, C., Marques da Silva, M., Kober, M., Wimmer, A., Dusing, J. and Hager, G., 2024. Laser-structured thin film sensor technology for sliding bearings in internal combustion engines. *Journal of Engineering for Gas Turbines and Power*. doi: 10.1115/1.4064451.
16. Birada, L.M., Pullela, V. and Thella, B.R., 2024. Triboinformatic machine learning model for frictional behavior and wear volume loss prediction of SS316L alloy clad with WC/NiCrBSi. *Journal of Bio- and Tribo-Corrosion*, 10(3), p.68. doi: 10.1007/s40735-024-00874-w.
17. Nowak, W.J. and Wierzba, B., 2018. Effect of surface treatment on high-temperature oxidation behavior of IN 713C. *Journal of Materials Engineering and Performance*. doi: 10.1007/s11665-018-3621-2.
18. Jaspert, D., Ebel, M., Eckhardt, A. and Poeppelbuss, J., 2021. Smart retrofitting in manufacturing: A systematic review. *Journal of Cleaner Production*. doi: 10.1016/j.jclepro.2021.127555.
19. Zhou, K., Liu, Z., Qiao, Y., Xiang, T. and Loy, C.C., 2023. Domain generalization: A survey. *IEEE Transactions on Pattern Analysis and Machine Intelligence*. doi: 10.1109/TPAMI.2022.3195549.
20. Marian, M. and Tremmel, S., 2023. Physics-informed machine learning—an emerging trend in tribology. *Lubricants*, 11(11), p.463. doi: 10.3390/lubricants11110463.
21. Igartua, A. *et al.*, 2020. Surface treatment solutions to green tribology. *Coatings*, 10(7), p.634. doi: 10.3390/coatings10070634.
22. Suryawanshi, S., Bhosale, D.G., Rukhande, S. and Vasudev, H., 2025. Advanced statistical modelling of tribological performance in Ni-based thermal spray coatings: A definitive screening approach. *Results in Surfaces and Interfaces*, 18, p.100435. doi: 10.1016/j.rsurfi.2025.100435.

23. Rukhande, S. *et al.*, 2025. High-temperature tribological performance of Ni-based coating on 316L stainless steel. *Results in Surfaces and Interfaces*, 18, p.100388. doi: 10.1016/j.rsurfi.2024.100388.
24. Verma, P.K., Vasudev, H. and Kumar, V., 2025. Dental implant materials and designs and failure during implant placement and loading: A review. In: *Advanced Manufacturing Processes*. CRC Press, pp.138–150.
25. Verma, P.K., Vasudev, H. and Kumar, V., 2025. Bioactivity and corrosion performance of low-velocity oxy-fuel-sprayed hydroxyapatite-based coatings on a surgical-grade stainless steel. *Transactions of the Indian Institute of Metals*, 78(8), p.185.
26. Verma, P.K., Vasudev, H. and Kumar, V., 2025. Effect of electro-corrosion behavior of different grades of stainless steel used in bio-implants: A review. In: *Advanced Manufacturing Processes*. CRC Press, pp.269–283.
27. Verma, P.K. and Vasudev, H., 2025. Importance of hydroxyapatite coatings in improving new bone osteogenesis: A review. In: *Advanced Manufacturing Processes*. CRC Press, pp.284–299.
28. Dutta, V., Thakur, L., Singh, B. and Vasudev, H., 2022. A study of erosion–corrosion behaviour of friction stir-processed chromium-reinforced NiAl bronze composite. *Materials*, 15(15), p.5401.
29. Singh, M., Vasudev, H. and Kumar, R., 2021. Corrosion and tribological behaviour of BN thin films deposited using magnetron sputtering. *International Journal of Surface Engineering and Interdisciplinary Materials Science (IJSEIMS)*, 9(2), pp.24–39.
30. Sharanabasava, H., Raviprakash, M., Prasad, C.D., Ramesh, M.R., Phanibhushana, M.V., Vasudev, H. and Kumar, S., 2023. Microstructure, mechanical and wear properties of SiC and Mo reinforced NiCr microwave cladding. *Advances in Materials and Processing Technologies*, 10, pp.3620–3633.
31. Singh, M., Vasudev, H. and Singh, M., 2022. Surface protection of SS-316L with boron nitride based thin films using radio frequency magnetron sputtering technique. *Journal of Electrochemical Science and Engineering*, 12(5), pp.851–863.
32. Prashar, G., Vasudev, H., Thakur, L. and Bansal, A., 2023. Performance of thermally sprayed hydroxyapatite coatings for biomedical implants: A comprehensive review. *Surface Review and Letters*, 30(01), p.2241001.
33. Vasudev, H., 2020. Wear characteristics of Ni-WC powder deposited by using a microwave route on mild steel: Microwave cladding of Ni-WC. *International Journal of Surface Engineering and Interdisciplinary Materials Science (IJSEIMS)*, 8(1), pp.44–54.
34. Parkash, J., Saggu, H.S. and Vasudev, H., 2022. A short review on the performance of high velocity oxy-fuel coatings in boiler steel applications. *Materials Today: Proceedings*, 50, pp.1442–1446.
35. Singh, P., Vasudev, H. and Bansal, A., 2024. Effect of post-heat treatment on the microstructural, mechanical, and bioactivity behavior of the microwave-assisted alumina-reinforced hydroxyapatite cladding. *Proceedings of the Institution of Mechanical Engineers, Part E: Journal of Process Mechanical Engineering*, 238(3), pp.1060–1071.
36. Verma, P.K., Vasudev, H., Singh, G. and Singh, S., 2024. A review on the microstructural and biomedical properties of implants manufactured using additive manufacturing. *Journal of The Institution of Engineers (India): Series D*, 106, pp.1–15.
37. Singh, S.P., Verma, P.K., Singh, S., Kapoor, M. and Balu, R., 2024. Microstructural characterization and electrochemical behaviour of HVOF sprayed WC-Co-Cr coatings on 316L boiler steel stainless steel. *Results in Surfaces and Interfaces*, 16, p.100248.
38. Verma, P.K., Vasudev, H. and Kumar, V., 2025. Biocompatibility and corrosion performance of HA-based LVOF sprayed and post-annealed coatings on SS 254 surgical-grade bio-implant material. *Ceramics International*, 51(10), pp.13207–13225.

39. Verma, P.K., Vasudev, H. and Singh, G., 2025. Improving the corrosion performance of SS 304 stainless steel using HVOF-sprayed WC-Co-Cr coatings in chloride environments. *Journal of Bio-and Tribo-Corrosion*, 11(4), pp.1–13.
40. Verma, P.K., Minhas, A.S., Singh, P. and Kumar, S., 2024, February. Slurry erosion behaviour of thermal sprayed Al2O3 and Cr2O3 coatings for turbine steels. AIP Conference Proceedings. AIP Publishing LLC, Vol. 2986, No. 1, p.020012.
41. Verma, P.K., Kumar, V. and Vasudev, H., 2024. Bioactivity and corrosion analysis of thermally sprayed hydroxyapatite based coatings. *Journal of Electrochemical Science and Engineering*, 14(4), pp.419–439.

9 Case Studies and Practical Real-World Examples in Smart Tribology

Praveen Kumar Verma, Hitesh Vasudev, and Shivam Kumar

9.1 INTRODUCTION

In exploring the possibilities offered by intelligent tribology within the cross-disciplinary domains of materials science and mechanical engineering, there is high promise. There is a higher integration of smart technologies within engineering and tribology (the study of friction, wear, lubrication, and preservation of mechanical systems). Digital tribology has moved out of the confines of laboratory systems into intelligent systems built and diagnosed to perform predictive and preventive maintenance in real time, self-adaptive coatings, and integrated smart self-adaptive coatings with embedded sensors and smart coatings. Intelligent predictive maintenance systems are developed and implemented within operational digital automatic maintenance frameworks of maintenance design engineering systems and collaboration between Siemens AG, the Schaeffler Group, and Honeywell International Inc., and are sustainably functionally designed within a circular economy. Advanced cross-sector predictive operational control designed within the operational redundancy of maintenance models of integrated smart materials and systems maintenance redundancies of tribological systems nest operational efficiencies within aerospace, energy, transportation, and manufacturing industries [1]. The described operational compliance frameworks include a case study detailing predictive maintenance integrated operational design systems.

9.2 SMART TRIBOLOGY IN INDUSTRIAL SYSTEMS: THE ROLE OF IOT AND PREDICTIVE ANALYTICS

To stay within seamless operational frameworks, various different types of machinery, along with each of their components, must adapt and respond to different load

 DOI: 10.1201/9781003741152-9

variations, operational speeds, and even extreme scenarios. Many traditional tribological methods, which deal with continuous oversight and preventative maintenance, neglect the impact of dynamic wear in modern systems. Smart tribology is the application of a digital interface to a contact interface. Siemens AG is a leader in integrating Internet of Things (IoT)-based monitoring systems and has digitalized condition monitoring, especially in critical tribology systems, such as turbines and compressors, which rotate and machine. Predictive maintenance by smart systems at Siemens combines wear monitoring with other systems of acoustic emission, differential thermal imaging, and condition monitoring of the machine. Information generated with monitoring systems and thermal imaging is foundational to Siemens MindSphere, a cloud-based echo system that uses machine learning to identify patterns of friction wear and lubricant aging. This approach operationally augments predictive maintenance and, by integrating it into the equipment maintenance strategy, improves the overall maintenance strategy. Staff can control the wear failure and optimize operational availability by managing the breakdown of seals, bearing assemblies, and leakages [2, 3].

In industries that rely on automation and their associated power plants, equipment malfunctions can lead to severe losses, making large profit-driven enterprises like Siemens' gas turbine fleets worth monitoring. Tribological monitoring allows Siemens to increase turbine maintenance intervals by 25% and decrease lubricant consumption by 18% in a maintenance cycle. This represents a clear improvement in operational sustainability. Siemens considers the innovative coatings and smart materials functioning with variable performance as smart tribology. Siemens has collaborated on turbine blade and shaft self-lubricating adaptive nanostructured coatings that are tribological variable performance materials that self-adjust to varying temperature and load differences. Various coatings with embedded nanosensors and friction-sensitive materials perform tribological temperature regulation and self-lubrication, enhancing real-time diagnostics. Siemens has adopted cycle maintenance in sustainable waste management, promoting circular predictive maintenance, component remanufacturing, and smart sustainable waste management linear with predictive maintenance [3].

9.3 DIGITALIZED PREDICTIVE MAINTENANCE AND SMART COATINGS

Siemens AG has utilized digital advancements in the innovation of tribology smart coatings. These coatings respond in real time to changing operational conditions. Siemens' smart coatings vary in lubricity and hardness based on operational temperature fluctuations in turbine and compressor systems. These temperature-responsive coatings are based on TiAlN and incorporate a thermoresponsive nano-coating feature. Such coatings actively monitor contact temperature, and withstand wear caused by fluctuating stress and friction coefficients during operational periods. These advanced coating systems incorporate thermoresponsive plating with embedded nano-scaled particulates and coatings. Integrated sensing systems record operational parameters and offer real-time wear measurement.

Digital twin technology has been integrated in predictive maintenance as well. A virtual copy that receives real-time data from IoT sensors is created for each mechanical part, be it a bearing or a gear assembly. The predictive maintenance solution enables engineers to accurately simulate wear behavior and forecast service intervals. Predictive analytics operationalizes predictive maintenance by cross-referencing operational and tribological data from analyzers to forecast optimal maintenance intervals. Integrated smart coatings, IoT-enabled predictive monitoring, and real-time analytics provide tangible results, including increased asset life, optimized lubricant consumption, and reduced unplanned downtimes. The synergistic use of digital and tribological technology empowers companies to achieve their objectives with respect to industrial ecosystem waste reduction, energy use, and waste management [4].

9.4 SMART BEARINGS, ADAPTIVE LUBRICATION, AND DATA-DRIVEN OPTIMIZATION

The system eliminates fixed scheduling of actuators, allowing the friction surface to be passively modified. The focus on friction loss quantification with the system, dynamic predictive systems, and interfacial friction adjustment in real time solidifies the relevance of interfacial predictive systems. The focus of Schaeffler's innovations is the AI-based condition monitoring systems and digital predictive adaptive interfacial friction adjustment maintenance systems called SmartCheck and OPTIME [5].

downtimes and costly failures by anticipating micro-pitting, Reliability requires the ability to monitor hundreds of bearings at once in industries like steel manufacturing, cement milling, and wind power generation. Schaeffler showed the use of distributed sensing networks for the real-time monitoring of tribological conditions of components and systems. Notably, a European steel mill fitted with OPTIME sensors recorded a 40% drop in unplanned downtime and a 30% decrease in lubrication waste in the first year post-installation. Maintenance strategies staunchly avoid expensive downtimes and costly failures by anticipating micro-pitting, preventing lubrication film breakdown, enabling non-disruptive targeted repairs, and maintaining components within predefined operational conditions through continuous condition monitoring. Moreover, maintenance process parameter optimization, guided by market feedback, minimizes downtime and reduces lubrication waste through an encapsulated, control-oriented optimization framework. These included load redistribution recalibration, lubricant viscosity adjustment, operational speed regulation, and other energy dissipation braking controls. Energy losses can be reduced during braking and other dissipation energy directives. All parts of OPTIME are coordinated with real-time fault diagnosis and vibration analysis conducted by the SmartCheck system. SmartCheck analyzes fault types by performing complete fault analysis through frequency domain signature analysis, measuring fault type temperature increases, and classifying fault types using embedded artificial intelligence. SmartCheck integrates with plant asset management (PAM) systems to monitor and control automated work order generation, maintenance prioritization, and scheduling. The operational synergy of SmartCheck enables alignment of the organizational strata with the maintenance and surveillance of tribological interfaces. This is indicative of the shift from reactive maintenance to an integrated system of predictive maintenance, which is predictive maintenance tribology. This is aimed at improving the predictive maintenance of the industrial system, operational productivity, and efficiency [6].

Intelligent tribology includes a Schaeffler research focus on hybrid ceramic bearing designs. In addition to other innovative materials engineered for high tribological performance, hybrid ceramic bearing tribo-coatings apply materials science. Schaeffler designs high-performance tribological bearings with Si_3N_4–ceramic rolling elements. Si_3N_4–ceramics improve bearing performance for tribological applications because of the increased bearing hardness, thermal stability, wear resistance, and significant reduction of friction and rolling fatigue. D-c.d. coatings maintain the performance improvements of the bearings by increasing wear resistance, scuffing resistance during boundary lubrication, and overall performance of the bearings through spray, PVD, or other advanced layer deposition methods [7].

Tribology is largely the result of the fusion of engineered materials and IoT-based control systems. Each unit utilizing predictive control triggers dynamically scheduled lubrication control for each of Schaeffler's lubricating tribological products. Systems like Schaeffler's Concept

support automated systems for maintaining film control lubricating tribological films sustaining systems, incorporating control systems for gauging the operational temperature, load, and speed of the controlled electro-mechanical operating systems. There is socio-economic value to adapted control of the operational life of the systems and significant reduction of waste lubricating oil through primary ESG and the circular economy.

The ability to generate digital twins from tribological data allows for the predictive optimization of bearings for various operational load scenarios. AI digital twins facilitate the predictive simulation of critical tribological components such as, but not limited to, wear and lubrication, surface fatigue, and other associated phenomena, thus making predictive assessment of almost any tribological behavior highly reliable. Shifting design factors to forecast tribological behavior and control factors significantly minimizes the time and cost of physical prototyping [7].

From an environmental perspective, Schaeffler seems to benefit from cutting-edge developments in intelligent tribology. The predictive lubrication and condition monitoring systems deployed across several sectors have resulted in a decline of 15%–20% in energy use, 30% in maintenance costs, and a decrease in CO_2 emissions and other economically quantifiable metrics. These results highlight the transformative role that predictive tribology can play in addressing the sustainability challenges of modern manufacturing. The integration of sensor technology, automation, advanced materials, and systems engineering to develop the smart tribology systems is a remarkable achievement for Schaeffler. With regard to digital monitoring, the closed-loop systems design in advanced mechanical engineering facilitates a fundamental transformation of tribological interfaces from passive mechanical wear to self-optimizing systems. The synergistic effect of OPTIME, SmartCheck, and adaptive coating systems constitutes a major advance in the industry. There is no question that intelligent tribology enables the design and engineering refinements the advanced manufacturing sector requires to achieve unprecedented sustainability and operational reliability in manufacturing [8].

9.5 PREDICTIVE TRIBOLOGY IN AEROSPACE AND ENERGY SYSTEMS

One of the world's leading firms within the aerospace, automation, and energy systems domains is Honeywell International Inc., which utilizes advanced smart tribology. Thanks to its optimization of advanced predictive tribological engineering, Honeywell was able to successfully converge into an Industry 4.0 operational ecosystem while addressing multidisciplinary challenges in reliability, engineering efficiency, and sustainable engineering. With advances in automation across different domains, the automation of industries has focused on predictive automation that captures, processes, and refines the output of automation systems, thus enhancing predictive automation while minimizing automation downtime. Predictive aerospace maintenance is the domain of the systems directorate within Honeywell. The design combines systems with triaxial vibration, temperature, and oil-quality sensors that tribologically monitor each engine, gearbox, and drive system of an aircraft individually. These sensors are configured for data streaming to both edge and main AI clouds for central processing on engineering fault pattern and thermal event temperature wear lubricant breakdown and thermal wear. These systems provide predictive maintenance alerting, thus maintenance scheduling is no longer calendar-driven, and maintenance teams are able to focus on the condition-based fatigue of the components [9].

Within the context of the HTF7000 series, Honeywell continues to benefit from the Intelligent Tribological Monitoring System (ITMS). Together, the ITMS and Health and Usage Monitoring Systems (HUMS) reduce unscheduled maintenance by 25% due to their ability to identify bearing failure, oil leaks, and misalignments. Better allocation of maintenance, replacements, and service life of component parts reduces operational costs and improves safety. Also, employing the predictive capabilities of HUMS, Honeywell can perform predictive maintenance on an entire fleet of aircraft. The system analyzes flight data, compares it with tribological data, and isolates systemic failures in the engine to improve the engine design. Apart from the propulsion for aerospace, tribological predictive analytics has been implemented in energy and in industrial automation to Honeywell's Experion Process Knowledge System (PKS). Experion is a hybrid computer–physical system and control platform in which tribological monitoring is integrated into actuators, compressors, pumps, and rotary systems, where it is embedded in a wide range of rotating equipment. In IoT monitoring, measurement is concerned with the stabilization of lubrication, surface, and friction parameters. The results from the measurement process are associated with process variables and are analyzed with predictive analytical tools. The smooth integration of tribological control and measurement systems permits the constant adjustment of system efficiencies. In the case of a compressor system, for instance, due to frictional heating of the system and the subsequent breakdown of lubricant viscosity, the control parameters of the system can adjust to lower flow and/or RPM to maintain operational efficiency and reduce wear, and do so with control algorithm parameter modifications. Such systems demonstrate smart tribometry by altering mechanical contact surfaces, and are, with closed stream data control, self-controlling [10].

Multifunctional tribological easy-to-sustain coatings and materials have led to progress in self-monitoring and self-adapting tribological systems. Thermoresponsive and piezoelectric, nano-engineered, and in situ sensing materials combine to form coatings developed in Honeywell's tribological research. These coatings sense and react to frictional heating and mechanical contact vibration. For example, piezoelectric thin films can actively monitor system wear and fatigue while producing mechanical signals when stress is applied. Likewise, a system is able to self-monitor by thermally reactive and electrically resistive or emissive coatings, eliminating the need for external sensors when the system's electrical resistance or emissivity changes. These materials have also outperformed passive protective wear systems and self-sustaining tribological systems. Due to Honeywell's seamless integration of futuristic designs in materials and sensors with advanced forecasting methodologies predicated on the organization's proprietary embeddings, the company employs long short-term memory (LSTM) machine learning algorithms, as well as Bayesian networks, to mine thousands of historical operational time bank intervals to generate forecasts on the anticipated trajectory of wear and the rate of oxidation of the contained lubricant. Given the algorithmic constraints on predicting the time to maintenance, Honeywell, in resource-constrained business environments, enhances the mean time between failures (MTBF) of operational aerospace and energy platforms when maintenance is provided only once the degradation probability crosses a threshold delta [11].

Predictive tribology systems decrease energy and CO_2 and lubricant waste. Customers, for instance, report draining active energy monitors installed in compressors and mid-scale industrial applications that capture 10%–15% energy efficiency gains and lesser overall maintenance-driven environmental degradation. Predictive maintenance, from analytical monitoring of active systems, further mitigates environmental degradation of closed logistics systems from unscheduled maintenance downtime. Honeywell's approach is in line with the increasing prioritization in the marketplace of digital systems being deployed as twin ecosystems. Advanced digital tribological systems stand as virtual bearing, seal, and actuator predictive maintenance technologies. These operational virtual tribological components, framed as operational failure predictors, embody stress concentration and thermal gradient loss of materials. The coupling of systems on material science with AI and analytics on friction to solve tribological challenges is what we define as intelligence. To a great extent, Honeywell demonstrates an ability to integrate forecasting tribology within complex, multidomain systems. Honeywell sustained an understanding of employing a tribological approach to design and operational systems, such as in HUMS, Experion PKS, and the intelligent coatings of the firm. The combination of adaptive materials with the IoT and AI analytics further enables the system to be more efficient and, in addition, more energy optimized, and therefore drives the transition to a more autonomously self-sustaining and self-regulating industrial ecosystem. Honeywell has made the first major step to understand the significance of smart tribology and the positive consequences it has on the environment and technological systems, especially in the aerospace and energy sectors [12].

Coating technologies allow engineers from diverse fields to analyze surfaces and structures from novel viewpoints. They include evaluations of surface roughness; assessments of mechanical integrity characteristics such as tensile, compressive, and flexural strengths; and improvements in the performance of solar energy systems through alterations of their optical and thermal properties. Coatings facilitate the clear depiction of fluid dynamics within intricate networks of channels and transitions, such as pipelines and elbows, that are difficult to understand. Functional coatings enable the anticipation of stress effects on structural integrity and performance in precision-engineered systems, such as ergonomically sensitive seating configurations. This prolongs the lifespan of the systems, elevates design quality, and diminishes errors in multiphysics simulations [13–30].

9.6 DISCUSSION AND CONCLUSION

The industrial case studies reinforce the importance of sustainability. The examples of smart tribology studied here further engineering sustainability by reducing frictional energy losses, improving the lifecycle of the machinery, and improving operations. Smart systems are implemented with low friction interfaces and operational materials and are designed within remanufacturing circular economies. Sustainable engineering is further complemented by adaptive coatings by Siemens allowing for the reapplication of the coatings and reuse of components, smart lubrication systems by Schaeffler ensuring welded components are lubricated, and Honeywell's predictive systems helping engineering sustainability by avoiding overly redundant engineering with excess replacement components. It is the singular integration of tribology and data science that this case study is based on. Industry predictive analytics in tribology are some of the most advanced, using neural Bayesian networks and digital twin simulations. With AI pattern recognition and predictive lubrication, and advanced digital twin technology, Siemens, Schaeffler, and Honeywell are leaders in tribology. These innovations have changed the philosophy of maintenance.

REFERENCES

1. Arjunan, A., Baroutaji, A., Robinson, J. and Olabi, A.G., 2021. Smart tribological coating. In: *Encyclopedia of Smart Materials*. Elsevier, pp.414–425.
2. Muratore, C., Clarke, D.R., Jones, J.G. and Voevodin, A.A., 2008. Smart tribological coatings with wear sensing capability. *Wear*, 265(5–6), pp.913–920.
3. Holmberg, K. and Erdemir, A., 2017. Influence of tribology on global energy consumption, costs and emissions. *Friction*, 5(3), pp.263–284.Holmberg, K. and Erdemir, A., 2019. Influence of tribology on global energy consumption, costs, and emissions. *Friction*, 7(1), pp.42–70.

4. Zemke, R., 1985. The Honeywell studies: How managers learn to manage. *Training*, 22(8), pp.46–51.
5. Verma, J. and Khanna, A.S., 2023. Digital advancements in smart materials design and multifunctional coating manufacturing. *Physics Open*, 14, p.100133.
6. Ciulli, E., 2024. Vastness of tribology research fields and their contribution to sustainable development. *Lubricants*, 12(2), p.33.
7. Islam, M.M., Fazle, A.B. and Prodhan, R.K., 2022. AI-based smart coating degradation detection for offshore structures. *American Journal of Advanced Technology and Engineering Solutions*, 2(4), pp.1–34.
8. Momber, A.W., Möller, T., Langenkämper, D., Nattkemper, T.W. and Brün, D., 2022. A digital twin concept for the prescriptive maintenance of protective coating systems on wind turbine structures. *Wind Engineering*, 46(3), pp.949–971.
9. Lee, J., Ni, J., Singh, J., Jiang, B., Azamfar, M. and Feng, J., 2020. Intelligent maintenance systems and predictive manufacturing. *Journal of Manufacturing Science and Engineering*, 142(11), p.110805.
10. Abdullah, M., Motayed, M.S.H. and Albaruni, M.A.S.I., 2025. Advancements in tribology and their impact on space systems: A review. *Journal of Bio-and Tribo-Corrosion*, 11(4), p.123.
11. Sose, A.T., Joshi, S.Y., Kunche, L.K., Wang, F. and Deshmukh, S.A., 2023. A review of recent advances and applications of machine learning in tribology. *Physical Chemistry Chemical Physics*, 25(6), pp.4408–4443.
12. Paturi, U.M.R., Palakurthy, S.T. and Reddy, N.S., 2023. The role of machine learning in tribology: A systematic review. *Archives of Computational Methods in Engineering*, 30(2), pp.1345–1397.
13. Verma, P.K., Vasudev, H. and Kumar, V., 2025. Dental implant materials and designs and failure during implant placement and loading: a review. In: *Advanced Manufacturing Processes*. Taylor & Francis Group, pp.138–150.
14. Verma, P.K., Vasudev, H. and Kumar, V., 2025. Bioactivity and corrosion performance of low-velocity oxy-fuel-sprayed hydroxyapatite-based coatings on a surgical-grade stainless steel. *Transactions of the Indian Institute of Metals*, 78(8), p.185.
15. Verma, P.K., Vasudev, H. and Kumar, V., 2025. Effect of electro-corrosion behavior of different grades of stainless steel used in bio-implants: A review. In: *Advanced Manufacturing Processes*, pp.269–283. https://doi.org/10.1201/9781003476238
16. Verma, P.K. and Vasudev, H., 2025. Importance of hydroxyapatite coatings in improving new bone osteogenesis: A review. In: *Advanced Manufacturing Processes*. CRC Press, pp.284–299.
17. Dutta, V., Thakur, L., Singh, B. and Vasudev, H., 2022. A study of erosion–corrosion behaviour of friction stir-processed chromium-reinforced NiAl bronze composite. *Materials*, 15(15), p.5401.
18. Singh, M., Vasudev, H. and Kumar, R., 2021. Corrosion and tribological behaviour of BN thin films deposited using magnetron sputtering. *International Journal of Surface Engineering and Interdisciplinary Materials Science (IJSEIMS)*, 9(2), pp.24–39.
19. Sharanabasava, H., Raviprakash, M., Prasad, C.D., Ramesh, M.R., Phanibhushana, M.V., Vasudev, H. and Kumar, S., 2023. Microstructure, mechanical and wear properties of SiC and Mo reinforced NiCr microwave cladding. *Advances in Materials and Processing Technologies*, 10, pp.3620–3633.
20. Singh, M., Vasudev, H. and Singh, M., 2022. Surface protection of SS-316L with boron nitride based thin films using radio frequency magnetron sputtering technique. *Journal of Electrochemical Science and Engineering*, 12(5), pp.851–863.
21. Prashar, G., Vasudev, H., Thakur, L. and Bansal, A., 2023. Performance of thermally sprayed hydroxyapatite coatings for biomedical implants: A comprehensive review. *Surface Review and Letters*, 30(01), p.2241001.

22. Vasudev, H., 2020. Wear characteristics of Ni-WC powder deposited by using a microwave route on mild steel: Microwave cladding of Ni-WC. *International Journal of Surface Engineering and Interdisciplinary Materials Science (IJSEIMS)*, 8(1), pp.44–54.
23. Parkash, J., Saggu, H.S. and Vasudev, H., 2022. A short review on the performance of high velocity oxy-fuel coatings in boiler steel applications. *Materials Today: Proceedings*, 50, pp.1442–1446.
24. Singh, P., Vasudev, H. and Bansal, A., 2024. Effect of post-heat treatment on the microstructural, mechanical, and bioactivity behavior of the microwave-assisted alumina-reinforced hydroxyapatite cladding. *Proceedings of the Institution of Mechanical Engineers, Part E: Journal of Process Mechanical Engineering*, 238(3), pp.1060–1071.
25. Verma, P.K., Kumar, V. and Vasudev, H., 2024. Bioactivity and corrosion analysis of thermally sprayed hydroxyapatite based coatings. *Journal of Electrochemical Science and Engineering*, 14(4), pp.419–439.
26. Verma, P.K., Vasudev, H. and Kumar, V., 2025. Biocompatibility and corrosion performance of HA-based LVOF sprayed and post-annealed coatings on SS 254 surgical-grade bio-implant material. *Ceramics International*, 51(10), pp.13207–13225.
27. Verma, P.K., Vasudev, H., Singh, G. and Singh, S., 2024. A review on the microstructural and biomedical properties of implants manufactured using additive manufacturing. *Journal of The Institution of Engineers (India): Series D*, 106, pp.1–15.
28. Singh, S.P., Verma, P.K., Singh, S., Kapoor, M. and Balu, R., 2024. Microstructural characterization and electrochemical behaviour of HVOF sprayed WC-Co-Cr coatings on 316L boiler steel stainless steel. *Results in Surfaces and Interfaces*, 16, p.100248.
29. Verma, P.K., Minhas, A.S., Singh, P. and Kumar, S., 2024, February. Slurry erosion behaviour of thermal sprayed Al2O3 and Cr2O3 coatings for turbine steels. AIP Conference Proceedings. AIP Publishing LLC, Vol. 2986, No. 1, p.020012.
30. Verma, P.K., Vasudev, H. and Singh, G., 2025. Improving the corrosion performance of SS 304 stainless steel using HVOF-sprayed WC-Co-Cr coatings in chloride environments. *Journal of Bio-and Tribo-Corrosion*, 11(4), pp.1–13.

10 Emerging Future Trends and Innovations in Tribology for the Age of Industry 4.0, and Future Outlook on Industry 5.0

Hitesh Vasudev and Gursimran Kaur

10.1 INTRODUCTION

The multidisciplinary science that studies friction, wear and lubrication is known as tribology and has traditionally been a cornerstone of mechanical engineering and materials science. With the development of industries moving toward more complex and computerized cyber-physical systems, tribology began playing a more extended role in maximizing surface interactions to realize system-level efficiency, sustainability, and reliability. The advent of Industry 4.0, which is also typified by smart factories, autonomous machines, ubiquitous connectivity, advanced analytics, and digital twins, has placed tribology as a key enabler of intelligent manufacturing and predictive decision-making. The change to data-driven, model-assisted, and sensor-supported tribological research, replacing the traditional form of empirical experimentation, can be seen as one of the most radical changes in the contemporary industrial environment. Under the definition of smart manufacturing, tribological interfaces no longer represent passive surfaces; instead, they present proactive actors to machine health, continuously measured via embedded sensors, Industrial Internet of Things (IIoT) architectures, edge cameras, and artificial intelligence (AI).

The growing complexity of engineered systems, including automotive parts and electric vehicles, aerospace structures, high-performance turbines, and biomedical implants, requires a greater understanding of the behavior of surfaces in dynamic, real-world operating environments. Conventional testing techniques will be inadequate to sustain the faster development schedule, budget limitations, and dependability requirements of Industry 4.0. As a result, new methods and approaches to the field of tribology, including tribo-informatics, physics-informed modeling, prediction based on machine learning, and the introduction of digital twins have become

DOI: 10.1201/9781003741152-10

very important tools in the modern world of tribology. Tribo-informatics is the combination of structured experimental data, microstructural characterization, and sophisticated data analytics that can reveal undiscovered trends and can be used to optimize surface engineering strategies in a manner that is not possible using conventional analysis. Research has enabled machine-learning algorithms to precisely determine material parameters such as friction coefficient and wear rate, eliminating the need to conduct a large-scale trial and error to develop a material, and making the iteration of material design faster [1, 2].

Multiscale and multiphysics modeling is also becoming more and more significant alongside data-centric approaches. These computational models span the gap between the surface interactions of atoms, the evolution of microstructure, the dynamics of lubrication, and the wear behavior on a macroscopic scale. High-level simulations, in turn, give valuable information that may be used to create realistic digital twins—computer models that develop in the real world using sensor data. With the growing dominance in industrial industries, the digital twin concept incorporates tribology-related features like the status of lubrication, temperature profile, vibration data, and surface wear, which become essential in predictive maintenance. The gear systems that have recently been developed on the digital twin have provided the example of how vibration-based wear prediction can greatly increase the reliability of key rotating parts [3, 4].

Tribology is important in strategy due to the fact that this branch of engineering is not restricted to predictive maintenance and optimization of machines. This smooth integration of tribological dynamics with smart sensors, automation, real-time control, and complex production processes in Industry 4.0 will change the way machines communicate, regulate themselves, and respond to additional environmental challenges. Next-generation industrial ecosystems are also adopting smart lubricants, self-adaptive surfaces, and stimuli-responsive surface materials, which turn out to be an intrinsic part of the industrial ecosystem. These sophisticated surfaces and lubrication systems have the ability to automatically adjust their characteristics depending on temperature, stress, contamination, or wear. The incorporation of those adaptive materials into cyber-physical settings of production can afford real-time control in tribological conditions, which would enhance the efficiency of energy, minimize downtime, and minimize the environmental impact.

With industries starting to move toward Industry 5.0, which focuses on human-centricity, sustainability, resilience, and the ethical incorporation of intelligent technologies, tribology expectations are changing. Whereas Industry 4.0 is mainly about automation and digital transformation, Industry 5.0 takes into account the needs of society, the welfare of workers, environmental responsibility, and the idea of the circular economy. Within this new environment, tribology will play an important role in machine performance as well as looking toward larger sustainability objectives including minimization of energy use using ultra-low-friction surfaces, generating biodegradable or recyclable lubricants, and the design of environmentally friendly coatings. Carbon emissions, resource use, and the component lifecycle can be significantly influenced by sustainable material development frameworks in the context of tribology; hence, complying with the Industry 5.0 environmental vision [5].

Moreover, the human–AI partnership, which is at the heart of Industry 5.0, will transform tribological systems to become explainable, transparent, and user-oriented in design. Such internalized tribological models will include decision-support systems that provide technicians, operators, and engineers with intuitive and augmented-reality displays of wear condition, lubrication requirements, and system health. This correspondence of the technological advancement to human skills encourages safer working environments, higher productivity, and reduced potential risks of operation. Tribology will be crucial in making precision in touch, reliability, and flexibility possible in areas like healthcare, aerospace, robotics, and energy, which are the primary focus of the human-centric vision of Industry 5.0.

10.2 INNOVATIONS IN TRIBOLOGY TOWARDS THE AGE OF INDUSTRY 4.0

Industry 4.0 has transformed the industrial ecosystem through the integration of cyber-physical systems, automation, AI, IIoT, and decision-making guided by data. These technologies have radically changed tribology as a discipline that was predominantly experimental and post-failure to one that is predictive, intelligent, and highly integrated in contemporary manufacturing. Bearings, gears, sliding contacts, seals, coatings, and lubricated systems are examples of tribological interfaces that have become active sources of data capable of self-monitoring, self-optimizing, and early predicting failures. With industries shifting toward modern smart predictive maintenance and real-time performance optimization through a conversion of traditional corrective maintenance, tribology has been a key enabling factor in reliability, energy efficiency, and system sustainability.

10.2.1 Data-Driven Tribology and Tribo-Informatics

The first wave of Industry 4.0-based innovation is the discovery of tribo-informatics, which is an interdisciplinary method that integrates tribological data, machine learning, microstructural characterization, and sophisticated statistical modeling. Tribo-informatics is designed to determine the latent wear behavior, frictional response, and lubrication mechanism patterns that have been inconspicuous in the traditional analysis. Machine-learned tribology has been used with success to forecast wear rate, coefficient of friction (COF), state of lubrication, and material degradation under different load speeds and environmental conditions.

It is shown by recent research that machine learning could exhibit the wear and friction behavior of composite materials with high accuracy so that optimal materials can be achieved quickly without the need to undergo numerous physical tests [6]. The additional studies indicate that the combination of microstructural aspects and ML algorithms can be used to improve the predictive ability of highly complex alloy systems to assist the designers in determining the most suitable process–experimental combinations to achieve a desired tribological performance [7]. Such advances constitute a paradigm shift: no longer do engineers have to resort to long-duration testing to shortlist potentially viable compositions, coatings, and process paths, but rather predictive models can be employed to do so.

Tribo-informatics can also be used to screen high-throughput new lubricants, solid film finishes, and reinforced composites. Neural networks, ensemble learning, and regression trees provide an opportunity to mine large databases of tribological experiments that were not previously fully utilized. This assists in the creation of new scientific information, performance optimization, and less time and cost of adopting it in an industry.

10.2.2 IIoT and Smart Sensors Tribological Monitoring

Smart sensors, IIoT devices, and embedded monitoring systems are another significant breakthrough of Industry 4.0 that is implemented in tribological interfaces. In real time, tribosystems can now be fitted with microsensors that are sensitive to vibration, temperature, acoustic emissions, the thickness of lubrication films, and wear debris. Such sensors produce real-time data streams that can be sent to cloud architectures or edge compute units to provide immediate processing.

Sensor-assisted tribology allows condition-based lubrication, i.e., the lubricant flow, viscosity, or additives are automatically controlled based on real-time system needs. Consequently, excessive lubrication, thermal breakdown of lubricants, and starvation can be avoided—increasing efficiency and component life.

Reliability of machine components is also enhanced by a significant margin when predictive algorithms are combined with IIoT sensors. It has been demonstrated that deep learning models can process vibration and acoustic indicators of gear systems and bearings to detect wear and micropitting at earlier stages before a failure event eventually takes place [8]. This technology saves money on downtime and increases the operational viability of the power generation sector and aerospace, mining, and intelligent car systems.

10.2.3 Tribological Systems and Digital Twins

One of the most radical Industry 4.0 innovations in tribology is the digital twin (virtual copies of real systems). A digital twin is constantly fed with real-time data provided by sensors and reacts to the behavior of a tribological interface in operating conditions. This allows the wear progress, lubrication failure, frictional heating, and surface roughness changes to be predicted correctly.

Compared to the traditional simulation, digital twins are highly effective due to the fact that they are dynamically updated using real field data. Digital twin models have proven to be incredibly accurate in predicting wear propagation and time to failure in rotating machines, like gears, so that they can plan proactive maintenance schedules [9]. Other than fault detection, digital twins are used to optimize lubrication regimes, surface topography, and operational parameters to increase reliability.

In addition, digital twins enable the use of what-if simulations, which enable engineers to determine the impact of design, load, materials, or lubricant chemistry changes on long-term performance. This shortens product development cycles, lessens the physical prototyping that has to take place, and enhances the decision-making authority of industrial maintenance teams.

10.2.4 Sleek Coatings and Surface Engineering of Smart Manufacturing

Further developments in tribology as part of Industry 4.0 are in the context of more advanced surface engineering, such as nanostructured surfaces, multilayer materials, textured surfaces by laser, additive manufactured textures, and adaptive/self-healing surfaces. Computational tools are now available to support technologies like plasma spraying, laser cladding, physical vapor deposition (PVD), chemical vapor deposition (CVD), sol–gel coatings, and microtexturing, which predict adhesion strength, wear resistance, and friction behavior.

As an illustration, coatings reinforced with nanoparticles are better in terms of resisting abrasive and adhesive wear as they are harder and more thermally stable. Machine learning models can be used to correlate the parameters of reinforcement percentage, deposition, and microstructural features with performance indicators such as wear rate and COF, accelerating the identification of next-generation coatings to be used in Industry 4.0 applications [6, 7].

Another type of innovation is a self-lubricating or self-healing coating, which is a surface containing microcapsules or tribochemical additives that release lubricants or create protective tribofilms when heated or activated by mechanical forces. These surfaces will increase the independence of tribological systems, which suits the philosophy of self-reliance of smart factories.

10.2.5 Intelligent Lubricants and AI Lubrication Systems

Nanolubricants enhance carrying capacity, cut down friction, and provide thermal stability under severe conditions. Predictive technology can formulate lubricant formulations that have the most optimal particle size, concentration, and functionalization.

Lubrication systems that are controlled by AI can automatically adjust lubricant viscosity, additive replenishment rate, pressure, and additive concentration depending on operational conditions. This smart control not only uses fewer resources but is also more energy efficient, as well as being a part of sustainability efforts related to Industry 5.0.

These models are the basis of the combination of tribology and digital twins, which will allow understanding the process of contact development, the formation of tribofilms, crack formation, and wear propagation in a deeper way. The importance of sophisticated modeling in alleviating the disparity between theoretical forecasts and tribosystem behavior in the real world is pointed out in recent works [10].

Collectively, these innovations emphasize the way Industry 4.0 has made tribology an intelligent, data-driven, and computationally augmented space, which is needed by current manufacturing ecosystems.

10.3 FUTURE OUTLOOK ON INDUSTRY 5.0

10.3.1 Sustainable, Human-Centric, and Resilient Tribology Systems

Industry 5.0 is an expansion in the technological base of Industry 4.0, as it focuses on the human-centricity, sustainability, and resiliency of the system, prioritizing people and the environment in industrial innovation. The tribology in this new paradigm is no longer associated with predictive maintenance and data-driven optimization but a field that will deal with performance, societal demands, ecological footprint, and long-term reliability. The future of tribological systems will be closely connected with bioinspired materials, circular design, and decentralized manufacturing concepts like hybrid additive–subtractive. Materials that are self-lubricating, self-healing, and self-adaptive, which can autonomously respond to stress, frictional heating, and surface degradation will be important to the life extension of components and also minimizing reliance on petroleum-based lubricants. New studies on sustainable surface engineering and more eco-friendly lubrication technologies show the promise of biodegradable ionic liquids, green nanoadditives, and lubricants based on water, which still retain high load-carrying capacity with low impact on the environment [11]. With Industry 5.0 focused on the carbon-neutral position and integrating digitally with the ecological aspect, tribologists will have to develop superior coatings and interfaces that use less energy, enhance the ability to recycle, and provide traceability throughout the entire lifecycle of industrial equipment.

One more significant change in terms of Industry 5.0 is the strengthening of human–AI cooperation when the operators, engineers, and decision-makers will collaborate with the cognitive machines and not be replaced by them. This requires tribological systems that can be explained and that can be transparent and understood by non-experts. State-of-the-art digital twins will become hybrid AI operating systems that incorporate physics models with interpretable AI algorithms, and give easy-to-understand visualizations of wear, lubrication conditions, and surface integrity. Research suggests that the combination of vibration, acoustics, thermal imaging, and spectroscopy will be employed in the future to create multimodal fingerprints of tribology in order to gain greater insight into the mechanisms of human contact at interfaces [12]. Wearable sensors and augmented reality (AR) interfaces will enable technicians to see the visualization of lubrication requirements, surface roughness variation, or an imminent failure in real-time, enhancing the quality of the working environment and accuracy of decisions. Also, a tough tribological architecture will get at the heart of the developing critical infrastructures: smart grid, self-driving transportation, aerospace propulsion, biomedical implants, and new-generation robotics. These applications need interfaces of friction to manage the unexpected dynamic loads, thermal shocks, or abrupt changes in the environment. The development of multiscale models and tribochemically engineered materials should enable the creation of interfaces that will be able to sustain extreme service conditions with low chances of failure [13]. All these human-oriented innovations will make tribology a convenient, adaptive, and eco-friendly field that is consistent with the ethical and social values of Industry 5.0.

10.3.2 Cognitive Tribology, Decentralized Manufacturing, and Intelligent Materials of Next-Generation Systems

Other perspectives of the future of tribology in the context of Industry 5.0 include the swift expansion of cognitive tribology whereby systems can learn, rationalize, and modify their frictional responses by utilizing real-time feedback. In this vision, tribological interfaces are not going to be mere passive devices but active cognitive entities in cyber-biological ecosystems. This shift is reflected by AI-controlled lubricants, which adjust viscosity in response to sensor feedback, materials that adjust their morphology when subjected to stress, and surfaces that grow protective tribofilms due to autonomous mechanochemical reactions. The promising outcomes of the current area of nano-engineered surfaces and multifunctional coatings are evidenced by the recent achievements of ultra-low friction, lower wear, and high temperatures in extreme operating conditions [14]. As additive manufacturing transitions to Industry 5.0, tribologists will take more advantage of 3D-printed metamaterials, graded surfaces, and textures based on topology optimization that are tailored at micro- and nanoscale to ensure that a given tribological objective is met. This customization is also in line with the personalization of Industry 5.0 and its flexible nature mainly in biomedical engineering, robotics, and electric mobility.

Besides material innovations, Industry 5.0 fosters the principles of decentralized, robust production, whereby small distributed manufacturing facilities demand highly dependable and self-sufficient tribological elements. Frictional interfaces will enable autonomous micro-factories to operate long without maintenance, which is enabled by self-regulating wear compensation and lubrication and AI-controlled wear compensation. The predictive digital ecosystems will combine tribology with larger sustainability analytics systems, making it possible to track the carbon footprint in real time, evaluate the degradation of lubricants, and predict the end-of-life circularity. With the help of the green transition of global industries, tribology integration with energy optimization, lifecycle engineering, and digital product passports will be additional sources of support [15]. The future perspective of Industry 5.0 is based on tribological systems, which are not only intelligent and high-performing but also ethical, environmentally conscious, and in harmony with the people who operate it and the natural ecosystems. This kind of transformation makes tribology an essential component of the next industrial revolution, which is the determinant of the reliability, sustainability, and societal contribution to technologies.

10.4 CONCLUSIONS

The high rate of Industry 4.0 and possible Industry 5.0 revolution is turning tribology into a more intelligent, networked, data-driven science. The recent developments in machine learning, IIoT, digital twins, and sustainable materials are transforming the way friction, wear, and lubrication are learned, forecast, and optimized in a wide range of engineering systems. The following conclusions have been drawn from this chapter:

1. Tribology is becoming a smart, data-driven field required for reliable, efficient, and predictive maintenance in modern Industry 4.0 manufacturing ecosystems.
2. Machine learning, tribo-informatics, and multiscale modeling are important in materials development, prediction of their performance, and optimization of frictional interfaces under complex industrial operating conditions a lot faster.
3. IIoT connectivity and smart sensors, proactive lubrication, early fault detection, and real-time monitoring in critical engineering applications have become possible, and this approach has proven to be resilient to operations.
4. Digital twins combine reaction-based modeling with real-time sensor feedback and provide high-fidelity, high-frequency predictions of wear, and tribological system health.
5. For the tribological components of the next generation, advanced wear resistance, sustainability, and versatility are offered by advanced coatings, self-healing materials, and nano-lubricants.
6. Industry 5.0 also focuses on human-centric sustainable tribology by using explicable AI tools, eco-friendly lubricants, and bioinspired surfaces to support safer and more environmentally friendly engineering systems.
7. Cognitive tribology and autonomous surface adaptation make possible frictional interfaces that may make intelligent decisions and optimize themselves to operate in harsh service conditions.
8. The future of tribology will combine environmental objectives, human cooperation, robust materials, and the digital ecosystem to support industrial sustainability and technological revolution.

REFERENCES

1. Pasha, M. B., Rao, R. N., Ismail, S., et al. (2024). Tribo-informatics approach to predict wear and friction coefficient of Mg/Si_3N_4 composites using machine learning techniques. *Tribology International*, *196*, 109696.
2. Sheikh, K. A., & Khan, M. M. (2024). Tribo-informatics analysis of in-situ TiC reinforced ZA27 alloy: Microstructural insights and wear performance modeling by machine learning. *Tribology International*, *199*, 110021.
3. Feng, K., et al. (2025). A digital twin methodology for vibration-based monitoring and predicting gear wear. *Wear*, *571*(1), 205806.
4. van Dinter, R., Tekinerdogan, B., & Catal, C. (2022). Predictive maintenance using digital twins: A systematic literature review. *Information and Software Technology*, *151*, 107008.
5. Pastewka, L., et al. (2025). Modeling in tribology: Recent advances, applications, and future perspectives. *Tribology International*, *218*, 111326.
6. Pasha, M. B., Rao, R. N., Ismail, S., et al. (2024). Tribo-informatics approach to predict wear and friction coefficient of Mg/Si_3N_4 composites using machine learning techniques. *Tribology International*, *196*, 109696.
7. Sheikh, K. A., & Khan, M. M. (2024). Tribo-informatics analysis of in-situ TiC reinforced ZA27 alloy: Microstructural insights and wear performance modeling by machine learning. *Tribology International*, *199*, 110021.

8. Feng, K., et al. (2025). A digital twin methodology for vibration-based monitoring and predicting gear wear. *Wear, 571*, 205806.
9. van Dinter, R., Tekinerdogan, B., & Catal, C. (2022). Predictive maintenance using digital twins: A systematic literature review. *Information and Software Technology, 151*, 107008.
10. Pastewka, L., et al. (2025). Modeling in tribology: Recent advances, applications, and future perspectives. *Tribology International, 218*, 111326.
11. Kalin, M. (2022). Environmentally friendly lubrication systems: Trends and challenges. *Tribology International, 174*, 107728.
12. Lyu, Y., et al. (2023). Advanced sensor fusion techniques for smart tribological monitoring. *Wear, 518*, 204686.
13. Masen, M. A. (2021). Multi-scale modeling in tribology: Toward predictive and resilient interface design. *Tribology International, 160*, 107033.
14. Erdemir, A., & Martin, J. M. (2020). Engineered tribofilms and ultra-low friction surfaces. *Tribology International, 146*, 106064.
15. Nosonovsky, M., & Bhushan, B. (2021). Green tribology and sustainable surface engineering: Future pathways. *Journal of Cleaner Production, 279*, 123745.

Index

For Product Safety Concerns and Information please contact our EU
representative GPSR@taylorandfrancis.com
Taylor & Francis Verlag GmbH, Kaufingerstraße 24, 80331 München, Germany

www.ingramcontent.com/pod-product-compliance
Ingram Content Group UK Ltd.
Pitfield, Milton Keynes, MK11 3LW, UK
UKHW020207200926
475702UK00004B/648
* 9 7 8 1 0 4 1 2 4 2 8 9 5 *